Nonmetallic Materials and Composites at Low Temperatures 3

CRYOGENIC MATERIALS SERIES

Nonmetallic Materials and Composites at Low Temperatures
Edited by A. F. Clark, R. P. Reed, and G. Hartwig

Filamentary A15 Superconductors
Edited by Masaki Suenaga and A. F. Clark

Nonmetallic Materials and Composites at Low Temperatures 2
Edited by Günther Hartwig and David Evans

Austenitic Steels at Low Temperatures
Edited by R. P. Reed and T. Horiuchi

Nonmetallic Materials and Composites at Low Temperatures 3
Edited by Günther Hartwig and David Evans

Nonmetallic Materials and Composites at Low Temperatures 3

Edited by

Günther Hartwig

Nuclear Research Center
Institute for Technical Physics
Karlsruhe, Federal Republic of Germany

and

David Evans

Rutherford Appleton Laboratory
Oxfordshire, England

Springer Science+Business Media, LLC

Library of Congress Cataloging in Publication Data

Main entry under title:

Nonmetallic materials and composites at low temperatures 3.

(Cryogenic materials series)
"Proceedings of the Third ICMC Symposium on Nonmetallic Materials and Composites at Low Temperatures, held August 23-24, 1984 in Heidelberg, Federal Republic of Germany."
Bibliography: p.
Includes index.
1. Nonmetallic materials—Thermal properties—Congresses. 2. Composite materials—Thermal properties—Congresses. 3. Materials at low temperatures—Congresses. I. Hartwig, Günther. II. Evans, David, 1940 Oct. 8- . III. ICMC Symposium on Nonmetallic Materials and Composites at Low Temperatures (3rd: 1984: Heidelberg, Germany) IV. Series.
TA418.95.N663 1985 620.1′1216 85-24337
ISBN 978-1-4899-2012-6 ISBN 978-1-4899-2010-2 (eBook)
DOI 10.1007/978-1-4899-2010-2

Proceedings of the Third ICMC Symposium on Nonmetallic Materials and Composites at Low Temperatures, held August 23-24, 1984, in Heidelberg, Federal Republic of Germany

Originally published by Plenum Press, New York in 1986
Softcover reprint of the hardcover 1st edition 1986

1984 INTERNATIONAL CRYOGENIC MATERIALS CONFERENCE BOARD

PREFACE

This was the third meeting in the series of special topical conferences on Non-Metallic materials at low temperatures. The first meeting was in Munich in 1978, the second in Geneva (1980) and so Heidelberg 1984 seemed an obvious time to review some of the hopes and objectives of the earlier meetings. It is also appropriate to consider the changing needs of the cryogenic community and how best the theory and practice of Non-metallic materials can be applied to suit this dynamic young science.

The aims and objectives of the International Cryogenic Materials Board in sponsoring this meeting remain the same. Namely, to provide a forum where practicing Engineers can meet with materials suppliers and researchers in an attempt to ensure that a real understanding exists between the two sides of the Cryogenic Materials Community. In this atmosphere, real problems can be addressed together with full discussions of tried and tested practical solutions. It is in this way that knowledge and confidence may grow hand in hand with the logical growth of the industry.

The first volume in this Non-metallic series noted that - "Cryogenics is an emerging technology, filled with promises." The second volume considered the 'delicately balanced relationship between knowledge and experience,' together with the hope that ..." year by year the application of non-metallic materials to the low temperature environment advances to the stage of a science". This, the third volume, reflects the urgent need that was so strongly evident in the panel discussion at the end of the conference. A need for further characterization of standard materials, and for formulators and materials scientists to ensure that well characterized composite materials are available to the cryogenic community.

It is clear that the science of cryogenic materials is still in its infancy but this latest volume, containing as it does a world wide collection of the latest works on the subject, continues the aims and objectives outlined in the previous volumes, progress through full understanding, detailed and frank discussion and careful application.

In many ways 1984 was an important year for the cryogenic industry and this truly international meeting was timely and of equal importance. Non-metallic materials have an important role to play in the rapidly expanding 'industrial' cryogenic world, and the world should know, with confidence, that the many aspects of this role are being researched and documented with care and enthusiasm.

David Evans
Günther Hartwig

CONTENTS

THERMAL EXPANSION OF NON-METALS

Guy K. White

CSIRO Division of Applied Physics
Sydney, 2070 Australia

ABSTRACT

The thermal expansion of a solid is determined by two factors, namely the internal stress generated by temperature change and the elastic stiffness. In many non-metallic crystals both these factors show considerable anisotropy which leads to expansion coefficients and associated Grüneisen parameters varying greatly with temperature and from one material to another. Experimental data extending down to 2 K are available for many materials of the common crystal structures such as rock-salt, fluorite, zincblende, rutile, caesium chloride, rare gas solids (f.c.c.). Data are scarce for most anisotropic crystals, mixed systems and composites. A brief review is given of the available experimental data for non-metallic crystals and some glasses and polymers.

INTRODUCTION

There are three thermal properties which are of practical interest in cryogenic design, namely heat conductivity, heat capacity (C) and thermal expansion. This paper reviews briefly (see Barron *et al*[1] for fuller treatment) what is known esperimentally of the expansion behaviour below room temperature of solids which are representative of common crystal structures, some glasses and polymers, and a few mixtures and composites.

Before considering the individual categories, there are general background features to remember:

1. Firstly, the volume coefficient of expansion, β, depends on the product of two factors: one is the anharmonicity or internal pressure P generated in a solid as the temperature T is changed and the other is the compressibility χ. That is

$$\beta = (dV/dT)_P/V = \chi_T(dP/dT)_V.$$

Other things being equal, soft materials expand more than hard ones, as shown by linear coefficient, α, in Figure 1.

2. Thermal expansion is a property (like heat capacity) which is in-

trinsic to the host lattice and therefore is relatively insensitive to trace impurities (say $\leqslant$ 1 mole percent) except at temperatures well below the Debye characteristic temperature, θ_D. At temperatures $T \leqslant \theta/10$, magnetic impurities or localised modes of heavy impurity atoms may have an energy per atom which is larger than that of the host lattice and so be significant. From an engineering aspect this is not important as the significant practical changes in dimensions occur at high temperatures.

3. The ratio β/C is rather constant for most materials at temperatures $T > \theta/2$ as Grüneisen observed in his classic experiments over sixty years ago.

Grüneisen introduced the dimensionless proportionality factor or anharmonicity parameter

$$\gamma = \beta V/\chi_T C_V = \beta V/\chi_S C_P$$

where $\chi_T(\chi_S)$ is isothermal (adiabatic) compressibility and $C_V(C_P)$ is heat capacity at constant volume (pressure) of a molar volume V. γ is sometimes erroneously called a constant. In fact it is a weighted average of the values $\gamma_i = -(V/\omega_i)(d\omega_i/dV)$ for individual lattice vibrational modes of frequency ω_i. Since γ_i for many non-metal crystals varies from -2 or -3 (for some transverse 'guitar-string' modes of vibration) to +3, the weighted average will vary with the relative excitation and importance of different modes as the temperature changes.

At elevated temperatures where all modes are excited we should expect $\gamma_\infty = \langle\gamma_i\rangle$ to be sensibly constant. This appears to be true and C_V and βB_T are also fairly constant at these higher temperatures ($B = 1/\chi$).

4. At the Cryogenic Materials Conference in 1983, I reviewed briefly[2] the expansion behaviour of metals and pointed out that excepting those metals with strong magnetic interactions, γ does not change more than 20% or so on cooling. This is because γ_i values do not vary so greatly in close packed systems. In such cases it can be useful for design calculations to assume $\gamma \approx$ constant and calculate changes in length ℓ or volume V using tabulated heat capacities or Debye Functions. Of course, if $\alpha(T)$ is known experimentally this is unnecessary.

As an example of this calculation for a non-metal, consider polycrystalline (isotropic) alumina for which γ only varies from ca 1.25 ($T \rightarrow 0$) to 1.35 (1000 K). Cooling α-alumina from 293 K down to 77 K would give a contraction of

$$\Delta\ell/\ell = \Delta V/3V = (1/3)\int_{77}^{293} \beta dT = (1/3)\int_{77}^{293} \gamma\, C_V \chi_T dT/V$$

$$\simeq (\gamma\chi_T/3V)\,[\,U_{293} - U_{77}\,].$$

U is the internal energy which is tabulated in terms of Debye functions, e.g. see American Institute of Physics Handbook, Third Edition, 1972, p.4-112. Using $\gamma = 1.3$, $\theta_D \approx 950$ K, $V = 25.6$ cm^3, $\chi_T = 4 \times 10^{-12}$Pa^{-1} leads to $\Delta\ell/\ell = 0.34 \times 10^{-6} \times [1850-20] = 620 \times 10^{-6}$. Compare this calculated value with the observed average[3] for sapphire of 628×10^{-6}. The agreement is so close as to be fortuitous.

5. As we shall see below, $\gamma(T)$ can vary greatly particularly for the

more open crystal structures which favour low-lying transverse accoustic modes having negative values of γ_i.

In the low-temperature limit, only long waves are excited and ultrasonic measurements at different pressures (or volumes) lead to the limiting value, γ_0. This may be calculated for cubic materials from the weighted average of the pressure derivatives of the principal elastic stiffnesses c_{11}, $c' = (c_{11}-c_{12})/2$ and c_{44} using

$$\gamma_i = -\frac{1}{6} + \frac{d\ell n c_i}{dP}/2\chi \text{ (see appendix}^1\text{)}.$$

For anisotropic crystals θ_0 may be calculated from the appropriate c_{ij} at two or more pressures (or volumes) and then differentiated numerically to give

$$\gamma_0 = - d\ \ell n\ \theta_0 / d\ \ell n\ V.$$

CLOSE-PACKED RARE GAS SOLIDS. (Coordination number CN = 12)

Solid neon, argon, krypton are not of much concern to most cryogenic engineers but are an example of close-packed non-metals. They show remarkably small variation in γ_i with either direction, polarisation or wave-number. For example, neutron scattering experiments on neon crystal show γ_i to all lie within the range 2.5 to 3.1 and the average $\gamma(T)$ from expansion data only varies between 2.58 and 2.7.

CAESIUM CHLORIDE STRUCTURE (CN = 8).

The caesium halides are also almost 'perfect' Grüneisen solids with $\gamma \approx 2.0$ from 2 to 300 K. This is a happy accident as they are elastically anisotropic and theoretical models indicate γ_i values ranging from 1 to 3.

ROCK SALT STRUCTURE (CN = 6)

Extensive measurements on alkali halides with this structure show γ_∞ lies between 1.4 and 1.7 for all but γ_0 drops from 1.65 for LiF to ca 1.0 for Na halides, 0.3 to K halides and 0 for Rb halides. This drop is associated with weakening of the c_{44}-type shear stiffness and the fact that γ_i for such modes becomes small and then negative for K and Rb-halides, i.e. they soften under pressure.

Figure 2 illustrates the behaviour of $\gamma(T)$. MgO also has this structure but γ is sensibly constant with value of 1.5.

FLUORITE STRUCTURE

Data for the alkaline earth fluorides, PbF_2 and ThO_2 give the picture of $\gamma(T)$ in Figure 3. In the progression from Ca to Sr to Ba, the ion size increases and associated with this is decrease in shear stiffness c' and softening of the mode, particularly in the (110) direction. The peak in γ for PbF_2 may arise from a nearly TA ferroelectric optic mode.

ZINCBLENDE STRUCTURE (CN = 4)

In this tetrahedrally-bonded family there are also low-frequency TA modes controlled by the shear stiffness c' which can soften under pressure. $\alpha(T)$ and $\gamma(T)$ become more negative as we grogress from group IV (Ge, Si) to III-V to II-VI and finally to a I-VII compound (CuCl) for which $\gamma_0 \sim -2$ (Fig. 4). The progressive change seems to be linked to the increasingly ionic nature and reduced angular rigidity of the bonding.

Cu_2O consists of two interpenetrating zincblende lattices sharing common oxygen atoms and shows even more extreme negative values of α and γ. The expansion coefficient α is negative from 0 to 270 K, reaching $\alpha_{min} = -3 \times 10^{-6}\ K^{-1}$ at 80 K.

WURTZITE STRUCTURE

These are also tetrahedrally bonded but have hexagonal symmetry and are therefore anisotropic. Limited thermal data for ZnO, Cds, ice and β-AgI all indicate that β and γ are negative at low temperatures.

ANISOTROPIC-AXIAL SYMMETRY

For crystals of hexagonal, rhombohedral or trigonal structure two linear coefficients are necessary to define the expansion. One is parallel to the symmetry axes ($\alpha_{||}$) and the other is normal to it ($\alpha_{\perp}$). These in turn depend on a product of a Grüneisen parameter and an elastic compliance s_{ij}

so $\quad \alpha_{||} = [\gamma_{||} s_{33} + 2 s_{13} \gamma_{\perp}]\ C_p/V$

and $\quad \alpha_{\perp} = [\gamma_{\perp}(s_{11} + s_{12}) + s_{13}\gamma_{||}]\ C_p/V.$

In either case, the second term in the square bracket can give rise to a negative expansion via cross-contraction, if the first term is small. The compliance s_{13} is normally negative.

Important examples are the layered materials, graphite and boron nitride, for which there is strong covalent bonding within layers and weak van der Waals forces between layers. This results in s_{33} being larger (soft) than $s_{11} + s_{12}$. The parallel coefficient $\alpha_{||}$ is large and positive. $\alpha_{\perp}$ is negative partly due to cross-contraction and partly from negative value of $\gamma_{\perp}$. Other examples are the chain-like solids, including polymers and tellurium and selenium. Here the weak forces are between chains so that $\alpha_{\perp}$ is large and positive. Strong bonding occurs along the chains so that s_{33} is small and $\alpha_{||}$ is negative, at least at low temperatures.

POLYMERS

A chain-like crystal of polyoxymethylene (POM), was measured down to 2 K and results were qualitatively similar to those for Se and Te : $\alpha_{\perp}$ is large and positive while $\alpha_{||}$ is negative below 100 K.

Recently, polyethylene has been measured down to 2 K for samples of varying crystallinity and degree of orientation[4]. Values for sample of 80% crystallinity and a draw ratio of 11 gave values for $\alpha_{||}$ and $\alpha_{\perp}$ which should approach those of the single crystal. As seen in Fig. 5, the $\alpha(T)$ curves are somewhat similar to those of Se and Te.

Choy[5] has reviewed the expansivity of oriented polymers giving data on PS, PMMA, POM, PVC etc, down to 100 K. Volume 13 of the Thermophysical Properties of Matter[6] gives much data for isotropic polymers including epoxy resins at temperatures above 70 or 80 K.

MINERALS AND LOW-EXPANSION CERAMICS

Most of the common minerals in the earth's mantle are silicates for which there is a marked lack of experimental knowledge at low temperatures.

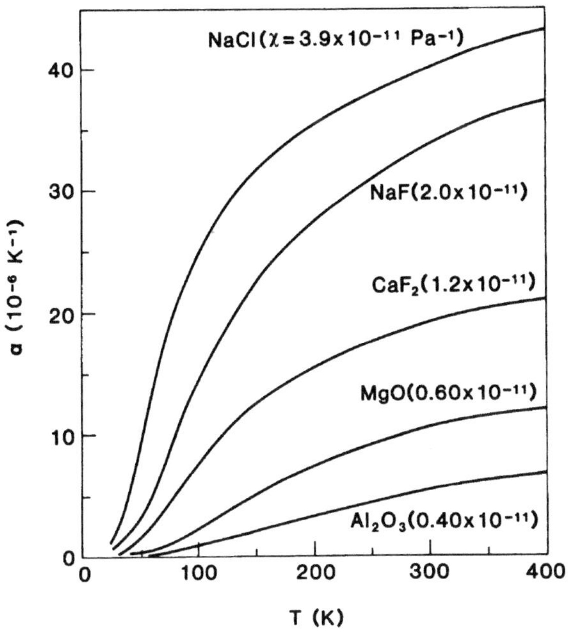

Fig. 1. α(T) for solids of different compressibility χ.

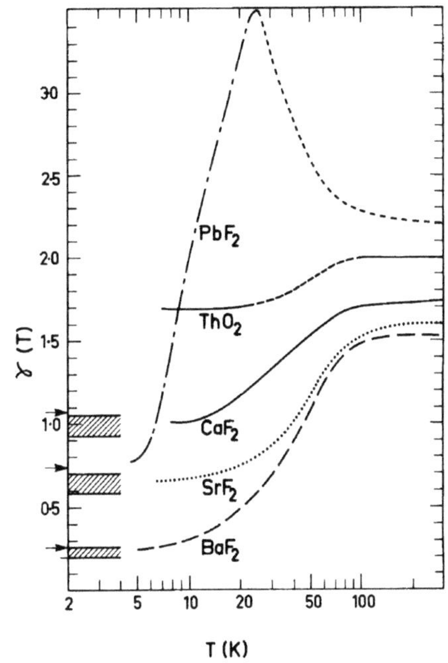

Fig. 3. γ(T) for some fluorites.

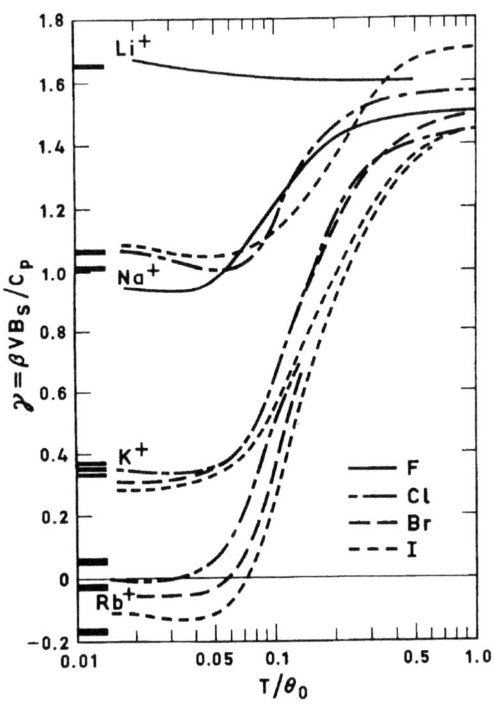

Fig. 2. γ(T) for alkali halides.

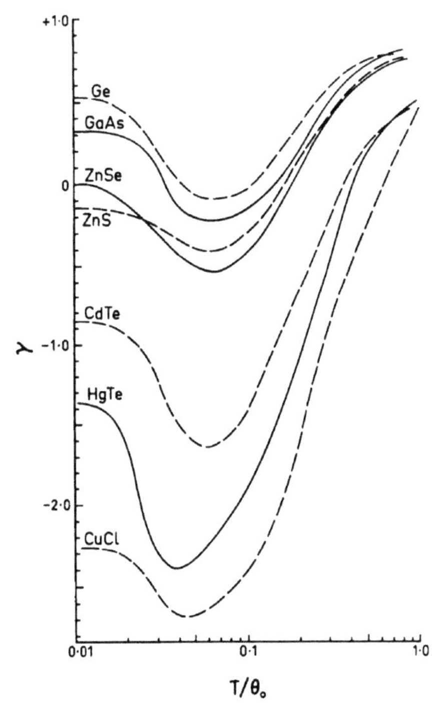

Fig. 4. γ(T) for zincblende structure.

An exception is α-quartz which consists of SiO_4 tetrahedra linked by shared oxygen atoms and having trigonal symmetry. The linear expansion is positive and large compared with vitreous silica, silicon or germanium. The coefficient increases markedly with T up to the α-β transition above which β-quartz has a small or negative expansion[7]. The expansion of α-quartz at normal temperatures is believed to be largely due to cooperative rotation of the tetrahedra and associated increase in the Si-O-Si angle.

Very-low-expansion materials are of technical importance both at low and high temperatures. The lithium alumino-silicates of β- spodumene, also described as substituted β- quartz structure, appear to have negative expansion behaviour and contribute to the 'zero-expansion' character of some ceramic glasses such as Zerodur[7] and CerVit.

Cordierite (2 MgO. $2A\ell_2O_3$. $5SiO_2$) is a aluminosilicate which has been measured at normal temperatures where α_c is negative and nearly twice the magnitude of α_a (positive) so that β is near zero[8].

Slack and Huseby[9] have discussed the expansion of some phenacite-type compounds including Zn_2SiO_4, Si_3N_4, Be_2SiO_4 and suggest that some of these might have negative expansions at low temperatures. They are open structures with tetrahedral bonding.

Related to the phenacite-structure is the olivine structure of Mg_2SiO_4 (forsterite) where the Mg ions are six-coordinated. The expansion of this orthorhombic mineral has been measured to low temperature with α_a, α_b and α_c all being positive[10]. The principal γ-values lie between 1.0 and 1.5 from 0 to 1000 K. At room temperature the average

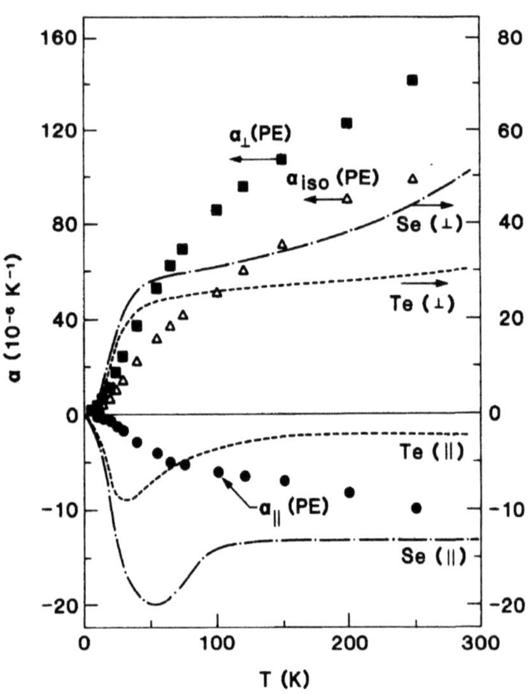

Fig. 5. α(T) for selenium, tellurium and polyethylene.

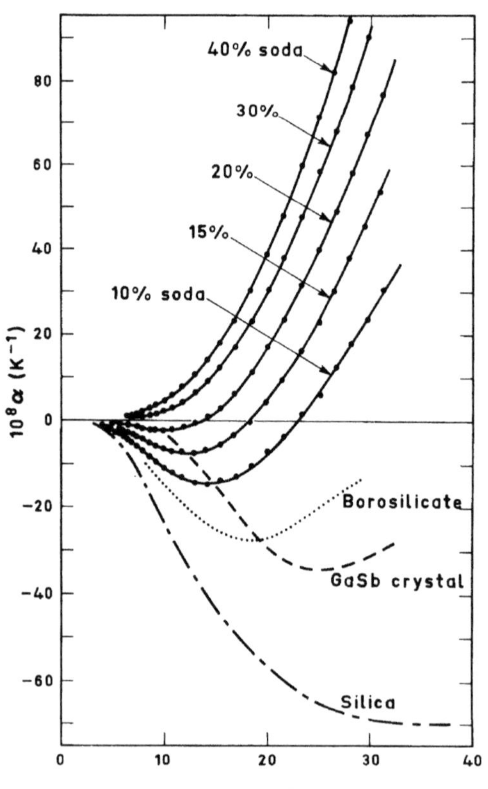

Fig. 6. α(T) for silicate glasses.

$\alpha \simeq 9 \times 10^{-6}\ K^{-1}$ compared with 12×10^{-6} for α- quartz and 10×10^{-6} for for MgO.

For the purposes of designing low-expansion mixtures or composites, it would be useful to have much more data at cryogenic temperatures on a variety of silicate minerals.

GLASSES

Glasses have been of practical interest for a very long time, but their thermal properties are still somewhat of a mystery. For example, vitreous silica, whose density is about 20% smaller than that of α-quartz, has an expansion coefficient which is negative below 150 K. The magnitude of α for SiO_2 depends on the thermal history (and density), being generally smaller in magnitude at high density. Other tetrahedrally bonded glasses such as GeO_2, BeF_2 also show negative expansion but of much smaller magnitude. Most non-tetrahedrally-bonded glasses, e.g. B_2O_3 and As_2S_3, have positive coefficients except perhaps below 2 K.

For SiO_2, additions of a few percent of network modifiers such as B_2O_3 (in Vycor) do not change this negative character significantly. Indeed the addition of 7.4 wt% TiO_2 which Corning use in their ULE silica enhances the negative α by about 30% below 50 K and yields a near-zero coefficient at room temperature.

However, network filling materials like soda, potash etc, reduce the negative component and so make α(T) much more positive (Fig. 6). Presumably the network filters inhibit those transverse modes which contribute to the contraction of the SiO_2 lattice.

MIXTURES AND AGGREGATES

There does not appear to be much data at low temperatures on well-defined mixed (2-component) systems which would allow comparison with theoretical models. Data on sintered mixtures of UO_2/ThO_2 at 283 K and 30 to 90 K indicate a roughly linear relation between α and relative concentration. The model of Turner[11] gives for an isotropic mixture

$$\alpha = \sum_i \alpha_i B_i V_i / \sum B_i V_i$$

where V_i is the relative volume occupied by the ith component and α_i and B_i are the corresponding expansion coefficient and bulk modulus.

Measurements on equimolar BaF_2/CaF_2 polycrystal (p. 662[1]) gave values from 300 down to 80 K which were much closer to BaF_2 than to CaF_2.

The analyses of effective thermal expansion coefficient of a polycrystalline aggregate of anisotropic constituent crystals is rather similar to the Voigt and Reuss averaging of eleastic moduli. Upper and lower bounds have been calculated recently by Hashin[12] for a number of metals but comparison with experiment is made difficult by the presence of preferred orientation in metals. For non-metals, the best data are probably for α-alumina. It is not particularly anisotropic (e.g. $\alpha_{||}/\alpha_{\perp} \approx 1.2$ at 200 K). The polycrystalline and the single crystal average ($= 2\alpha_{\perp} + \alpha_{||}/3$) agree within experimental error of ± 1%[3].

Technically important are the composites comtaining one or more crystal phases in an isotropic matrix. Calculations have been made by Levin, Hashin, Rosen[12], but well-defined experimental systems do not seem to have been measured. Yates and collaborators[13] have measured a number of graphite-epoxy systems which are of interest but are not readily comparable with theoretical models. It seems desirable to have good data on, for example, randomly distributed anisotropic particles (of known expansion) in an isotropic matrix (of known expansion), also high-density compacts of isotropic particles having very different α-values would be of interest.

ACKNOWLEDGMENT

I thank my colleague, John Collins, for his frequent and generous assistance.

REFERENCES

1. T.H.K. Barron, J.G. Collins and G.K. White, Thermal expansion of solids at low temperatures, Adv. Phys., 29: 609-730 (1980).

2. G.K. White, Metals and Alloys : Expansion and contraction in "Advances in Cryogenic Engineering Materials" Vol. 30, Plenum Press, New York (1984) in press.

3. G.K. White and R.B. Roberts, Thermal expansion of reference materials : tungsten and α-$A\ell_2O_3$. High Temp-High Press 15 : 321-328 (1983).

4. G.K. White and C.L. Choy, Thermal expansion and Gruneisen parameter of isotropic and oriented polyethylene. J. Polym. Sci.:Phys. Sci. Ed. 22 : p.835-846 (1984).

5. C.L. Choy, Thermal expansivity of oriented polymers, Developments in Oriented Polymers -1. I.M. Ward, ed. Applied Science Publishers Ltd, Essex. pp. 121-151.

6. Y.S. Touloukian, R.K. Kirby, R.E. Taylor and T.Y.R. Lee, Thermophysical Properties of Matter, Vol. 13 - Thermal Expansion, Non-Metallic Solids, Plenum Press, New York (1977).

7. R.B. Roberts, R.J. Tainsh, G.K. White, Thermal properties of Zerodur at low temperatures, Cryogenics, 22 : 566-568 (1982).

8. D.L. Evans, G.R. Fischer, J.E. Geiger and F.W. Martin, Thermal Expansions and Chemical Modifications of Cordierite, J. Amer. Ceram. Soc. 63 : 601-728 (1980).

9. G.A. Slack and I.C. Huseby, Thermal Cruneisen parameters of phenacite-type compounds. J. Appl. Phys., 53 : 6817-6822 (1982).

10. G.K. White, R.B. Roberts and J.G. Collins, Thermal Properties and Grüneisen parameter of Mg_2SiO_4 (forsterite), High Temp-High Press, to be published (1985).

11. P.S. Turner, Thermal expansion stresses in reinforced plastics, J. Res. Nat. Bur. Stand. 37 : 239-250 (1946).

12. Z. Hashin, Thermal expansion of polycrystalline aggregates, I. Exact Analyses. J. Mech. Phys. Solids, 32 : 149-157 (1984).

13. S.J. Feltham, B. Yates and R.J. Martin, Thermal expansion of particulate-reinforced composites. J. Mat. Sci. 17 : 2309-2323 (1982).

THE LOW TEMPERATURE THERMAL CONDUCTIVITY OF RUBBER

J.J. Freeman and D. Greig

Department of Physics
University of Leeds
Leeds U.K.

ABSTRACT

The low temperature thermal conductivity, κ, of nitrile and natural rubber has been measured, together with the influences on κ of stretching and 'filling' with Carbon Black. These new results, as well as those presented previously, are used in an attempt to establish a qualitative relationship between the observed properties and polymer structure.

INTRODUCTION

The low temperature thermal conductivity, κ, of polymers has been the subject of much study during recent years[1,2]. These studies have included the effect on the magnitude and temperature dependence of κ due to (i) varying degrees of crystallinity, and (ii) orientation of the polymer. In a previous paper[3] we discussed some of the work already performed on amorphous and semicrystalline polymers, and highlighted our interest in measuring the thermal conductivity of amorphous rubber. In particular, we noticed in the literature that for most semicrystalline polymers, the thermal conductivity above $\sim$30K of stretched or extruded material increases significantly with increasing values of stretch ratio, λ, whilst virtually no change is noticed in the conductivity of amorphous polymers[4]. Results published on rubber however, show both of these tendencies and are therefore apparently contradictory[5,6,7]. Other than in reference 3, all previous measurements on rubber have been made at room temperature and above and have been made using various types of rubber. Since natural rubber crystallizes to a certain degree on both cooling and stretching, we also decided to investigate the effect on κ of crystallinity, which we have simulated in our amorphous (nitrile) rubber by adding Carbon Black. In this latest paper we report on the variation of κ with various degrees of filling with Carbon Black as well as on further investigations into the magnitude and temperature dependence of κ for various degrees of stretching.

All conductivity measurements were made between 1K and 100K using a steady state differential method, and carried out in a Helium-4 cryostat Details of the measuring procedure may be found in reference 3.

RESULTS AND DISCUSSION

The temperature dependence of κ for natural and nitrile rubber is shown in figure 1. At temperatures ∿100 K the magnitude of the conductivities are very similar, but at lower temperatures the conductivity of natural rubber falls well below that for nitrile rubber. As already mentioned, natural rubber has a tendency to crystallize on cooling whereas nitrile rubber does not. This is in accordance with the temperature dependence of κ which exhibits a "plateau region" between 5 and 20K, characteristic of all amorphus materials[8]. All further measurements were made using nitrile rubber (polybutadene-co-acronitrile) in which the random sequence of the monomer units present inhibits crystallization[9].

Effect of stretching

Figure 2 shows results obtained for macroscopic stretch ratios of 1·47 and 2·04. These results are similar to those for polyethylene[10], but must be interpreted differently since polyethylene is a semicrystalline polymer, and nitrile rubber is amorphous. For temperatures above ∿5K there is an increase in conductivity above the isotropic value by a factor of rather less than λ. This increase we postulate as being the result of straightening and orientating the chains in the direction of stretching, producing more efficient heat transfer and decreasing the number of scattering centres per unit length. Below ∿5K, the phonon wavelength and mean free path are longer and this effect becomes relatively unimportant.

Extrapolation of the data up to room temperature would give a similar but smaller increase in κ to that observed by Tautz[5] for natural gum rubber. Nevertheless, since natural rubber crystallizes on stretching, the larger increase in Tautz' results is only to be expected. It could be argued that the extremes of temperature together with stretching may cause some crystallization to occur. However, the effect of crystallization over the temperature range measured would be to lower κ rather than to raise it.

Figure 2 also shows results for a sample which has been repeatedly stretched and then held at an extension ratio of 1·72. The conductivity of this sample can be seen to be comparable to that of the unstretched

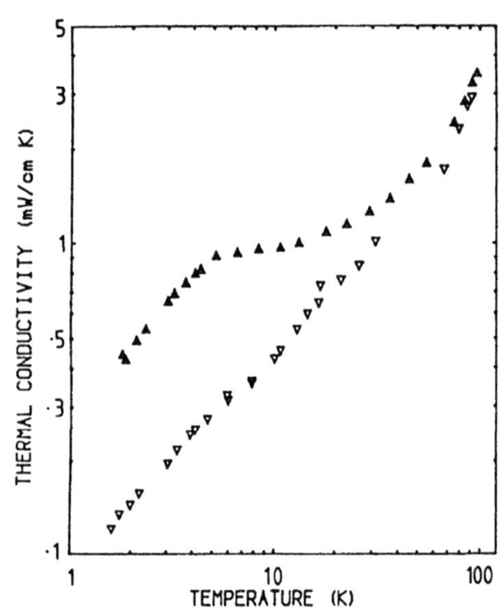

Fig. 1. Thermal conductivity of natural (∇) and nitrile (Δ) rubber

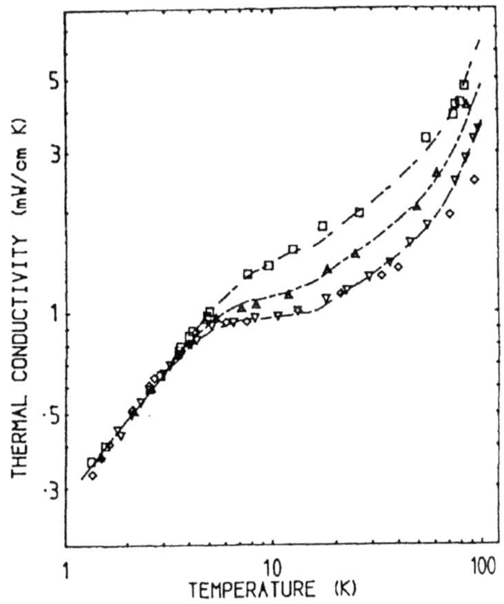

Fig. 2. Thermal conductivity of stretched nitrile rubber. Results for unstretched rubber (∇) and for extensions to $\lambda = 1{\cdot}47$ (Δ), $\lambda = 2{\cdot}04$ (□) and repeated extension to $\lambda = 1{\cdot}72$ (◇).

rubber. We suggest that this is an example of the " Mullins effect" [11]; that is, a smaller increase in the conductivity of a stretched material due to internal structural changes. This reduction in κ upon repeated stretching has also been observed by other workers[7].

An important distinction between amorphous rubber and other amorphous polymers at room temperature is that rubber is above its glass transition temperature, and the molecules can therefore freely rotate. Taking this into account, it seems reasonable that a greater increase in conductivity should be observed. However repeated stretching alters the structure of the rubber creating creep and the breaking of crosslinks, and hence reducing the rubber to the isotropic state.

Effect of filling with Carbon Black

The temperature dependence of κ for the three specimens used are shown in figure 3, together with the unfilled sample. The added concentrations are; 20%, 40%, and 60% by mass. The results are very similar to those for semicrystalline polymers such as polyethylene terephthalate[13], and for voids in ceramics[14]. At high temperatures (>40K) the magnitude of κ for all four samples is similar, but with the highest filler concentration having the highest conductivity. Below ∿40K the curves cross and separate with the most filled sample having the lowest conductivity.

Carbon Black MT N990 (Medium Thermal) is a crystalline form of graphite in which the Basal plane is folded in upon itself to form spherical particles of ∿0·25 - 0·35μm in diameter. This form of Carbon Black does not form aggregates. Carbon Black is an active filler for nitrile rubber; that is, the elastomer interacts (chemically and physically) very strongly with the filler surface. This is due to the polar nature of nitrile rubber - namely the presence of OH and COOH dipoles[9]. It is therefore postulated that due to the tight binding of the polymer to the filler, voids are unlikely to form (at least under non-stretching conditions).

The simplest explanation of the observed behaviour is that above ∿40K the filler has a higher conductivity κ_f, than that of the rubber matrix, κ_m, and that this enhances the conductivity of the bulk material. At ∿40K the conductivities κ_m and κ_f are comparable, while below

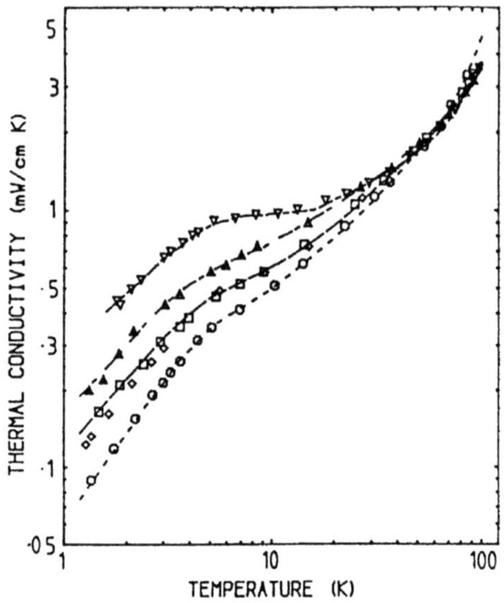

Fig. 3. Thermal conductivity of nitrile rubber filled with Carbon Black MT N990. Unfilled rubber (∇), and rubber filled with mass fractions 20% (Δ), 40% first set (□), 40% second set (◇), and 60% (O).

$\sim$40 κ_f falls well below that of κ_m. This implies that κ_f has a large temperature dependence which seems reasonable for crystalline material. (For crystalline solids below $\theta_D/30$, the conductivity is limited by boundary scattering and $\kappa \propto T^3$.)

Towards the bottom of the temperature range the curves for the filled material become nearly parallel with a temperature dependence which is linear. This dependence is characteristic of filled or semi-crystalline material[15,16] and is due to surface scattering from the impurity/matrix interface. Turk and Klemens[17] predict that in this region

$$\kappa \propto \frac{1}{V_f\, d} \qquad (1)$$

where V_f is the volume fraction of the crystalline phase, and d the diameter of the particles. The fraction V_f may be calculated using the known densities of the constituent materials. Our only unknown was the density of Carbon Black, which we were able to calculate using weighing techniques. We found a value that agreed to within 5% of valves quoted in literature. Between $\sim$1 and 5K we then find excellent proportionality between κ and $1/V_f$ as predicted by (1).

To test repeatability of our samples, a new sample filled to 40% was made. This second set of measurements is also displayed in fig.4 (40% 1 and 2), and shows excellent agreement with the previous data.

Effect of filling and stretching

Figure 4 shows the results obtained for stretching the 20% filled sample to twice its natural length. Above $\sim$30K the magnitude of κ is larger than that of the unstretched sample, but not by such a large factor as in the case of unfilled rubber. Below $\sim$30K, κ is actually reduced(by a factor of 2 at 2K). This overall relative decrease in κ is most easily attributed to the creation of voids around or within the carbon particles due to stress concentration. At higher temperatures (>7K) there will be a decrease in the number of connecting paths through the Carbon Black. At lower temperatures, both V_f and d are likely to increase, hence lowering κ whilst still keeping the observed linear temperature dependence.

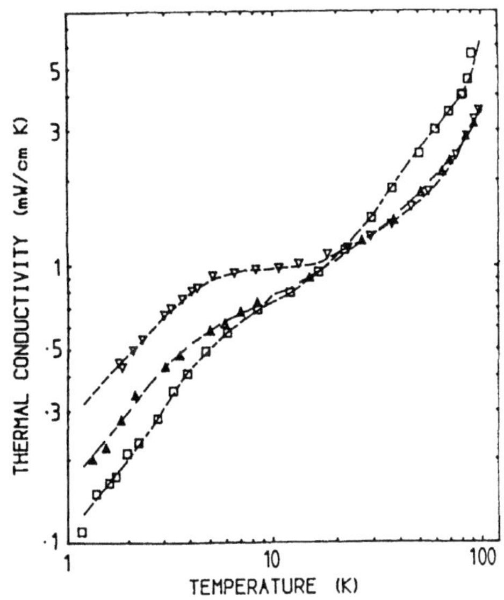

Fig. 4. Thermal conductivity of nitrile rubber filled with 20% (by mass) Carbon Black MT N990 with λ = 1 (Δ), λ ∿ 2 (□), together with unfilled unstretched rubber (∇).

ACKNOWLEDGEMENTS

We should like to thank Dr D. Hands of R.A.P.R.A. for supplying the specimens and for his advice and interest in this project and to acknowledge our appreciation to the S.E.R.C. for their continued financial support.

REFERENCES

1. C.L. Choy, Polymer 18: 984 (1977).
2. D. Greig, "Developments in Oriented Polymers - I", I.M. Ward, ed., Applied Science Publishers Ltd., London (1982).
3. J.J. Freeman and D. Greig, Advances in Cryogenic Engineering (Materials), Volume 30 (1983)
4. K.H. Hellewege, J. Hennig and W. Knappe, Kolloid ZZ Polym. 188: 121 (1963).
5. H. Tautz, Exper.Tech.der Phys. 7: 1 (1959).
6. D. Hands, Rubber Chem.Technol. 53: 80 (1980)
7. M. Pietrella, Colloid & Polymer Sci. 259: 111 (1981).
8. R.B. Stephens, Phys.Rev.B 8: 2896 (1973).
9. M. Morton, ed. Rubber Technology 2nd edn. Ch.1 (1973)
10. C.L. Choy and K. Young, Polymer 18: 769 (1977)
11. J.A.C. Harwood, L. Mullins and A.R. Payne, Polymer Letters 3: 1119 (1965).
12. M. Morton, ed. Rubber Technology, Ch. 3
13. C.L. Choy and D. Greig, J.Phys.C. 8: 3121 (1975)
14. E.P. Roth and A.C. Anderson, J.Appl.Phys., 47: 3644 (1976).
15. A.C. Bhattacharyya and A.C. Anderson, J.Low Temp.Phys. 35: 641 (1979).
16. D. Greig and M.S. Sahota, J.Phys.C 16: L1051 (1983).
17. L.A. Turk and P.G. Klemens, Phys.Rev.B 9: 4422 (1974).

THE THERMAL CONDUCTIVITY OF POLYMERS BELOW 1K

D. Greig and M. Sahota

Department of Physics
University of Leeds
Leeds U.K.

ABSTRACT

It has been well-established for a number of years that the thermal conductivity, κ, of all amorphous materials - including amorphous polymers - exhibits "universal" behaviour below 1K, falling as T^2 with magnitudes that are very similar between one material and the next. For semicrystalline polymers, on the other hand, the behaviour is quite different. At 1K the values of κ are an order of magnitude lower than in amorphous materials, with a temperature dependence in which κ is proportional to T. In this report we summarise recent measurements of κ below 1K, with emphasis on the effects of (i) changes in the degree of crystallinity and (ii) extrusion. Very generally κ tends to the universal T^2 dependence at very low temperatures (below 100mK), and when either the crystallinity is small or the extrusion ratio is large. We shall also report on measurements on a single crystal polymer, polydiacetylene, in which, as expected for a crystalline solid, the temperature dependence of κ approaches T^3.

INTRODUCTION

It has been well-established for a number of years that the low temperature thermal conductivities (κ) of all amorphous solids are very much alike, varying as T^2 below 1K with absolute values that are surprisingly similar over a wide range of different materials[1]. This general statement applies equally to inorganic glasses and to polymers. For semicrystalline polymers, on the other hand, the situation is quite different. Even in the normal "helium" range from 1 to 4K both the magnitude and temperature dependence of κ can vary appreciably, depending on the crystallinity and, to a lesser extent on the degree of extrusion of the material. A summary of the major features of this temperature range was presented at a previous ICMC meeting[2], and in a more comprehensive review published in 1982[3]. During the last two years the range of measurements has been extended to well below 1K, and in this paper we review these new developments. Here again the data are critically dependent on crystallinity and extrusion ratio, and we shall discuss each of these features in turn.

VARIATION WITH CRYSTALLINITY

Any change in the fractional crystallinity (X) has a dramatic effect on κ with values in semicrystalline polymers at 1K of up to an order of magnitude lower than the "universal" value in amorphous solids referred to above. As the temperature is lowered below 1K the thermal conductivity of partially crystalline material appears to vary as T, with no immediate evidence of the T^2 dependence that is characteristic of amorphous solids in general[4].

To investigate what happens at lower temperatures still we chose to make measurements on the semicrystalline polymer, polyethylene terephthalate (PET) - a polymer in which X can be varied by annealing from zero to about 0.5[5]. The measurements, reproduced in Figure 1 show that the linear temperature dependence occurs over a relatively limited range and when T is sufficiently low the various sets of measurements tend to converge. (The two most highly crystalline specimens were annealed at rather higher temperatures than the rest, leading to thicker crystallites and a possible explanation as to why, for them, the absolute values of κ remain rather low.)

The most significant feature of the results is that by 60mK the conductivity of all specimens varies as T^2. At that temperature the wavelength of the dominant phonons is estimated to be about 300nm - a value considerably greater than the thickness of the crystallites. The phonons no longer "see" the crystallites as anything other than point defects, and the more resistive 2-level scattering processes in the amorphous regions become dominant giving rise to the T^2 variation as observed.

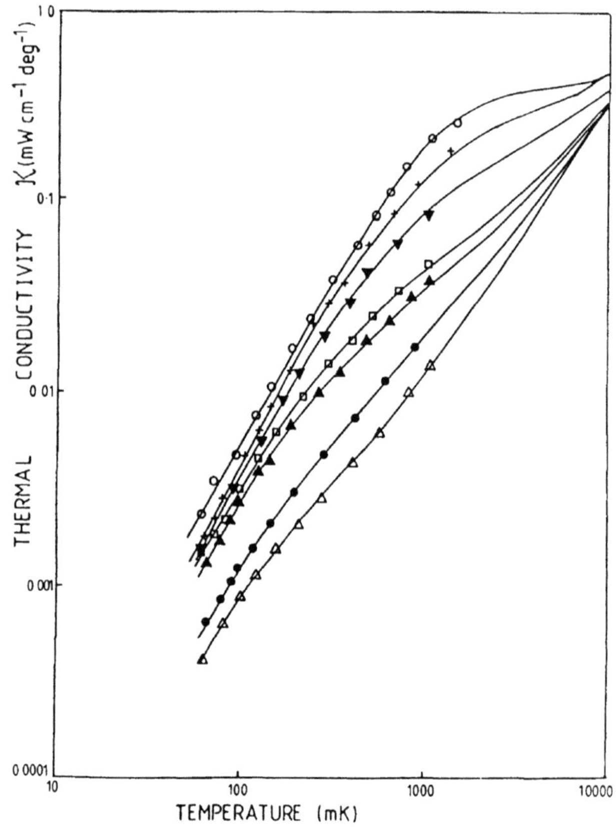

Fig. 1. Measured thermal conductivities of PET. The crystallinities are; O, 0%; +, 9%; ▼, 17%; □ 25%; ▲, 29%; ●, 39%; Δ, 51%.(Reproduced from reference 5.)

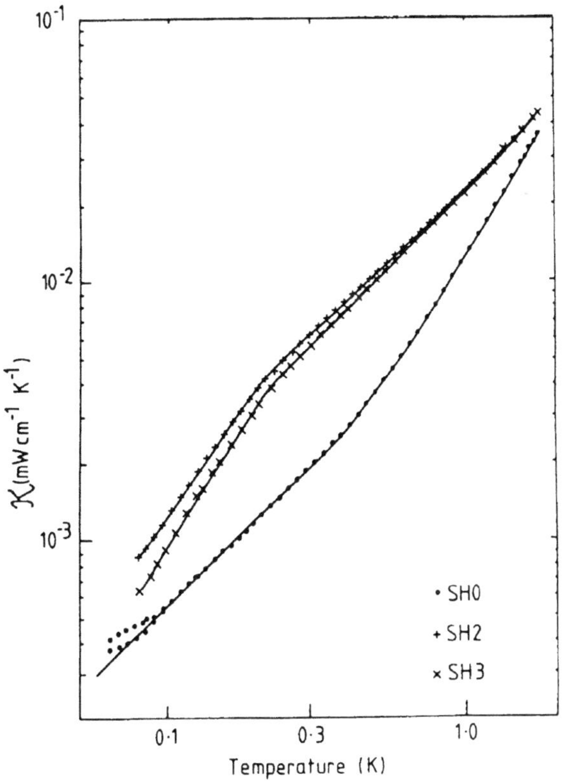

Fig. 2. Measured thermal conductivities of isotropic polyethylene (SH0), and extruded polyethylene with extrusion ratios 3·9(SH2) and 4·4(SH4). The latter measurements are made in the extrusion direction. (Reproduced from reference 8.)

VARIATION WITH EXTRUSION

In contrast to the above, drawing or extrusion has a surprisingly small effect on the values of κ of semicrystalline polymers in the normal helium range. This is also in contrast to the very striking change at all temperatures above about 25K (including room temperature) where κ in the extrusion direction increases in simple proportion to the degree of drawing or extrusion[6]. The implication is that below ∿25K the orientation of the crystallites brought about by the extrusion process no longer appears to matter.

Around 1K, however, the influence of extrusion re-emerges. We have mentioned in previous papers[4,7] how, for semicrystalline polymers, the slope of the κ versus T curves always undergoes a sharp decrease at a temperature T^*, typically of order 1K, and this we have associated with the dominant phonon wavelength becoming longer than the thickness of the crystallites. We have also pointed out that for measurements of κ in the direction of extrusion the value of T^* tends to increase (a) with the degree of extrusion and (b) as the extruding temperature is reduced. This is entirely in accordance with the observation that both of these processes give rise to an average separation between crystallites that is relatively small, suggesting that the crystalline structure is more "broken up".

This change in T^* implies that at temperatures well below 1K there should again be a significant variation in κ with degree of extrusion, and this Finlayson and Mason have recently confirmed[8] (Figure 2).

They have found that at 250mK the increase of conductivity in the extrusion direction in two samples of polyethylene of extrusion ratios 3·9 and 4·4 is roughly in proportion to the degree of extrusion, while at the same time they have shown that there is virtually no change in specific heat. However, below 200mK, there is a sharp increase in the slope of the κ v T curves for the extruded specimens, so that by the lowest measured temperatures (80mK) the spread in the values of κ has fallen from a factor of 4 to a factor of less than 2. Here again, at low enough temperatures, the results converge.

Very generally, then, we may say that the effect of extrusion on the thermal conductivity below 1K is to bring the measurements closer to the "universal" values of amorphous materials. We have also observed this trend in a series of specimens of extruded polypropylene with extrusion ratios of up to 10·5, where the data include measurements of κ perpendicular to the extrusion direction. It is interesting to note that at these temperatures the parallel and perpendicular conductivities are both greater than the isotropic values.

CRYSTALLINE POLYMERS

Finally we should like to make brief mention of the results of a collaborative experiment between Queen Mary College, London, G.E.C., and ourselves in measuring κ in the single crystal polymer, polydiacetylene[9]. At 1K the value of κ is about an order of magnitude greater than in amorphous polymers - and therefore 2 orders of magnitude greater than in a semicrystalline polymer - and varies rather slowly as $T^{\frac{1}{2}}$ between 1 and 50K. Below 1K, however, κ falls rapidly with the characteristic T^3 variation of crystalline solids. The value of the corresponding scattering rate (10^7 s^{-1}) is an order of magnitude greater than we might expect from the dimensions of the sample, and we have suggested that the "boundaries" are the gaps formed by cleavage during polymerisation. From the observed relaxation rate the average separation of the gaps is estimated to be about 100μm.

ACKNOWLEDGEMENTS

We are most grateful to Professor I.M. Ward and his colleagues for supplying specimens, to Dr D.M. Finlayson for reading the manuscript and for giving permission to use figure 2, and to the S.E.R.C. for continued financial support.

REFERENCES

1. R.B. Stephens, Low temperature specific heat and thermal conductivity of noncrystalline dielectric solids, Phys.Rev.B. 8: 2896-2904 (1973).
2. D. Greig and N.D. Hardy, Thermal conductivity at low temperatures in semicrystalline polymers, in: "Non-metallic Materials and Composites at Low Temperatures - 2", G. Hartwig and D. Evans, eds., Plenum Press, New York (1982), pp. 27-35.
3. D. Greig, Thermal conduction in oriented polymers, in: "Developments in Oriented Polymers - 1", I.M. Ward, ed., Applied Science Publishers Ltd., Barking (1982), pp. 79-120.
4. D.M. Finlayson, P. Mason, J.N. Rogers, and D. Greig, The thermal conductivity and specific heat of extruded polyethylene below 1K, J.Phys.C:Solid State Physics 13: L185-L188 (1980).
5. D. Greig and M. Sahota, The thermal conductivity of an amorphous material below 1K - variation with degree of crystallinity, J.Phys.C:Solid State Physics 16: L1051-L1054 (1983).
6. A.G. Gibson, D. Greig, M. Sahota, I.M. Ward and C.L. Choy, Thermal conductivity of ultrahigh-modulus polyethylene, J.Polym.Science, Polymer Letters Edn. 15: 183-192 (1977).
7. D. Greig and N.D. Hardy, The thermal conductivity of semicrystalline polymers at very low temperatures, J.de Physique 43: Colloque C6-69-71 (1982).

8. D.M. Finlayson and P. Mason, Thermal properties of a semicrystalline polymer at very low temperatures, J.Phys.C:Solid State Physics (1984) in the press.
9. M.N. Wybourne, B.J. Kiff, D.N. Batchelder, D. Greig, and M. Sahota, Acoustic-optical phonon scattering in the thermal conductivity of polydiacetylene single crystals, submitted to J.Phys. C:Solid State Physics (1984).

MECHANICAL PROPERTIES OF THREE CANDIDATE ORGANIC INSULATOR MATERIALS FOR FUSION REACTORS

D.S. Tucker, G.F. Hurley, and J.C. Kennedy

Los Alamos National Laboratory

Los Alamos, New Mexico, U.S.A

ABSTRACT

Insulators in superconducting magnets for fusion reactors will be exposed to radiation doses of 10 to 100 Mgy with more than half the dose coming from fast neutrons. One polyimide- and two epoxy-based laminates* were neutron irradiated to total doses of 2.6 and 4.1 x 10^{21} n/m^2, E>0.1 MeV at 4.2 K in the Intense Pulsed Neutron Source at Argonne National Laboratory. Flexural properties were determined at 75 K and compared to results from previous studies of gamma irradiation effects at Oak Ridge National Laboratory.

INTRODUCTION

Organic insulation has been specified for use in fusion reactor superconducting coils because of lower material and fabrication costs compared to other forms of insulation. Glass filled organics (epoxy and polyimide resins) are prime candidates since they provide not only electrical insulation, but also the mechanical strength necessary to withstand the high magnetic forces present during coil operation. However, these materials exhibit a greater sensitivity to irradiation than most other inorganic reactor components. The properties of organic materials and glasses are degraded by both neutron and gamma irradiation present in fusion reactors. Therefore, the radiation response of organic insulation can depend upon the sum of neutron and gamma-ray intensities which differ for various reactor designs.[1-4] In contrast, important magnet components such as the superconductor and the copper or aluminum stabilizer are sensitive only to neutron irradiation. Structural degradation of organics

---

* NP-530 - Polyimide resin matrix, analogue of G-10CR; Norplex Division, VOP Inc.

G-10CR - National Electrical Manufactures Association (NEMA). A heat-activated amine-catalyzed bisphenol solid-type epoxy resin laminate reinforced with continous filament E-glass fabric, silane finished; designated for cryogenic use; Spaulding Fibre Company.

G-11CR - Same as G-10CR, except uses an aromatic amine curing agent has also been shown to occur at lower neutron doses than those for the other classes of magnet materials.[5]

A number of studies have been carried out in recent years in which degradation of candidate epoxy and polyimide based fiberglass insulation and of other organic materials has been characterized by measurement of physical property changes after reactor irradiation at 4.2 K. Most of the imposed dose in those studies has been deposited by gamma irradiation. The results have shown that the post-irradiation mechanical properties of the high strength epoxy based materials are seriously degraded at doses less than 10^9 rads, while the polyimide based materials do not appear to show serious strength losses in the 10^9 to 10^{10} rad dose range.[6-12] Post-irradiation electrical properties do not exhibit as severe a degradation as do the mechanical properties in the 10^9 to 10^{10} rad dose range for most of the epoxy and polyimide based materials.[13]

Since a substantial part (50-90%) of the absorbed dose will result from energy deposited from fast neutrons, it is of the utmost importance to understand the effects of neutron irradiation on candidate organic insulator materials. It is also important to establish similarities and differences between radiation damage due to gamma-rays and fast neutrons, in order to account for spectral effects. Nishijima and Okada[14] compared the effects of gamma-rays and fast neutrons on the breaking strength and strain of epoxy resin and found that more damage was produced by a smaller dose of neutrons (10^6 rads) than by the gamma-rays (4×10^7 rads). This led them to conclude that simulation of neutron irradiation by substituting gamma-rays is not a safe practice.

In the present study we have determined post-irradiation mechanical properties (3 point bend) of two epoxy and one polyimide based glass reinforced materials. These materials were irradiated up to a neutron fluence of 4.1×10^{21} n/m^2, E>0.1 MeV, at 4.2 K and tested at 75 K. Results are also compared with results from previous studies by Coltman et. al. using gamma irradiation.[12]

EXPERIMENTAL

Samples for this study included G-10CR, G-11CR, and NP-530*. The irradiation was carried out in two parts.

Samples from a previous irradiation in the Intense Pulsed Neutron Source (IPNS-I) along with fresh samples were irradiated twice in the IPNS-II at Argonne National Laboratory using the following procedure. Samples were weighed, measured, photograghed, and then stacked in an aluminum foil pouch inside a copper cylinder. Nickel dosimetry wires were attached axially around the edge of the pouch at ninety degree intervals with an additional wire approximately in the center. Dosimetry from the nickel wires indicated fairly uniform doses of 1.3×10^{21} n/m^2, E>0.1 MeV for each irradiation. The gamma dose is not accurately known but is estimated to be 15% of that from neutrons. An exposure history is given in Table I.

Mechanical flexure tests were carried out at 75 K using a three point bend test fixture actuated by a cryogenic compression test apparatus described elsewhere.[14] Deflections were measured by means of an Linear Variable Differential Transformer (LVDT), which detected relative movement of the platens of the compression fixture. Length between outer spans was 25.4 mm, and pin diameter was 6.4 mm. Samples were 3.2 x 1.6 mm in cross section and were tested with the major stress parallel to the warp direction. Deflection rate was 1.3 mm/min. This test procedure follows the requirements of ASTM D790-80.

RESULTS

Test results were analyzed to determine modulus of rupture (MOR), flexural modulus (E), and failure strain (E_f). MOR was calculated using

the simple beam formula; $S=3PL/2bd^2$, where;
S = stress in the outer fibers at midspan, N/m^2 (psi),
P = load at a given point on the load deflection curve, N (lbf),
L = support span, m (in.),
b = width of beam tested, m (in.),
d = depth of beam tested, m (in.).
This formula is valid for maximum fiber strains of 5% or less, with flexural strength (MOR) being equal to the maximum stress in the outer fibers at the moment of break.

MOR results, figure 1, show that only G-10CR exhibited a significant loss in strength (67%) at a fluence of 4.1×10^{21} n/m^2. G-11CR showed only a small reduction in strength (12%), while NP-530 exhibited a small increase in flexural strength at this fluence.

The failure strain, defined as the nominal strain at fracture, is given in figure 2. The G-10CR samples show a decrease from an original value of 4.4% to approximately 1.85% at a fluence of 4.1×10^{21} n/m^2. Although the G-11CR control samples have a lower E_f value than the G-10CR samples, the decrease after irradiation is much less with a value of 3.8% at a fluence of 1.5×10^{21} n/m^2. After exhibiting an initial decrease to approximately 3% at a fluence of 1.5×10^{21} n/m^2, E_f values for the NP-530 samples remained constant.

Flexural modulus, figure 3, for all three materials remains relatively constant for all irradiation values. After an initial rise at a fluence of 1.5×10^{21} n/m^2, E for G-11CR decreases to the initial value (control sample) and remains constant. Only G-10CR shows a steady decrease in E, while NP-530 actually exhibits an increase in E for fluences greater than 1.5×10^{21} n/m^2.

DISCUSSION

Of the three candidate materials, G-10CR showed the reduction in strength with G-11CR exhibiting only a slight decrease and NP-530 actually displaying a slight increase at fluences greater than 2.6×10^{21} n/m^2. This behaviour can possibly be attributed to two factors. First, neutron irradiation dose is strongly sensitive to hydrogen content since at least one-half of the maximum energy can be transferred to hydrogen during elastic collision producing energetic recoil nuclei. The ionizing dose decreases in going from G-10CR to G-11CR to NP-530 as a result of the same trend in hydrogen content, which reflects in the strength results. The second factor, thought to be of greater importance, takes into account the nature of the chemical and molecular structure of both the resin materials and curing agents. Both G-10CR and G-11CR are identical except that G-11CR utilizes an aromatic cure, while G-10CR uses an aliphatic curing agent. It has been established that the resonant ring structure is more resistant to radiation than the straight chain aliphatic structure, due to the ability of the non-localized pi-electrons in the ring to absorb and dissipate energy more readily than a single bond

TABLE I. Exposure history of test samples

Neutron Fluence (10^{21} n/m^2, E>0.1 MeV)				
	IPNS-1	IPNS-2A	IPNS-2B	Cumulative
IPNS-I+IPNS-II	1.5	1.3	1.3	4.1
IPNS-II Only		1.3	1.3	2.6

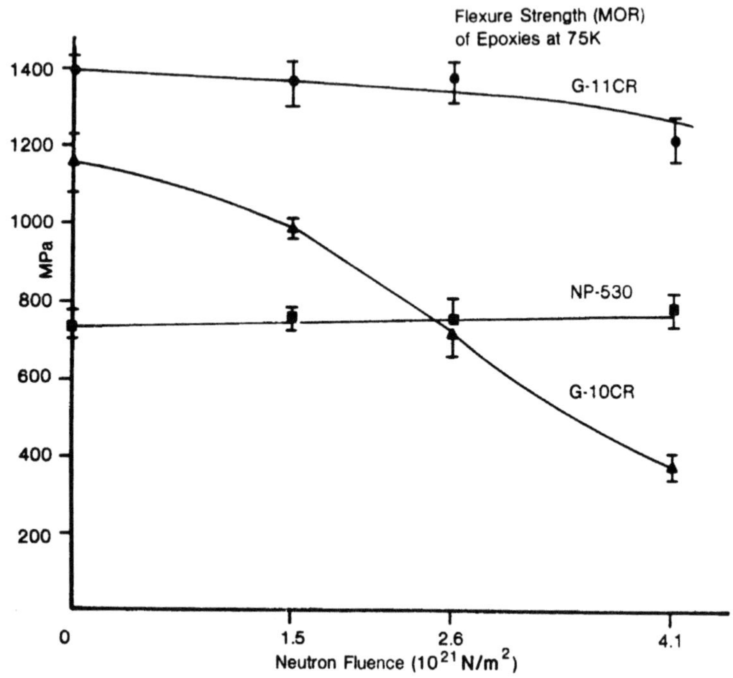

Fig. 1. Plot of MOR versus neutron fluence.

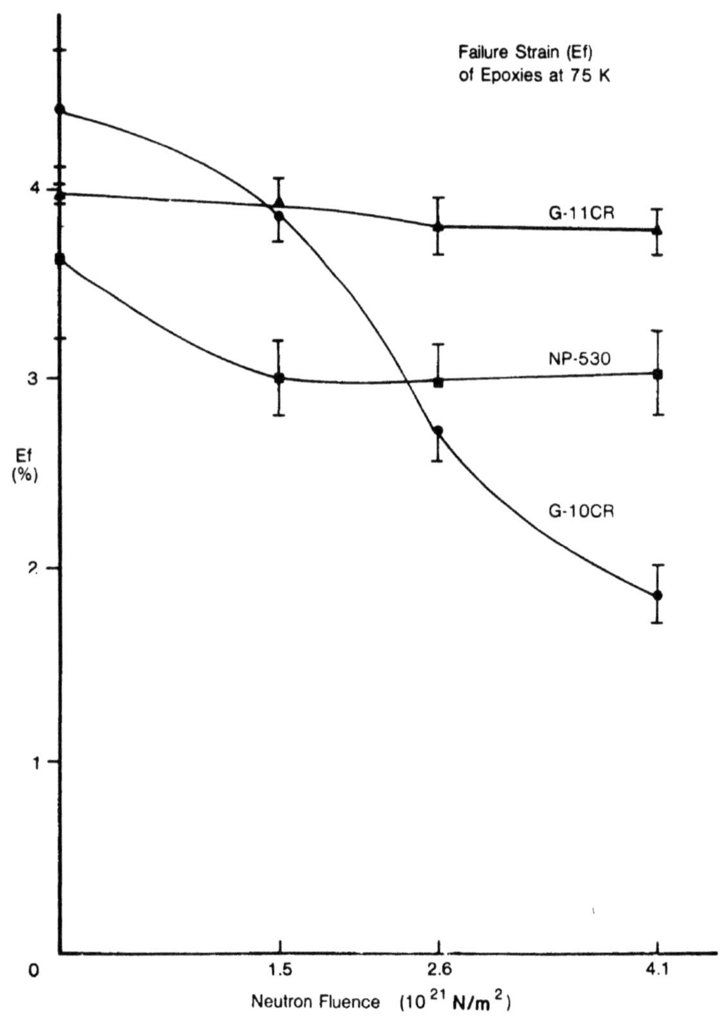

Fig. 2. Plot of failure strain vs. fluence

characteristic of the aliphatic structure.[15] The bond strength of the aromatic structure is also greater than that of the aliphatic structure which enhances radiation resistance.

NP-530 is a polyimide, which is a nitrogen carrying, condensed aromatic ring system. The ring is an electron-withdrawing group which makes possible resonance between nitrogen and the ring. This, in turn, increases radiation resistance. This structure also contains sidechains with double bonds present, which again increases the resonant energy and provides paths for energy to pass into and be evenly distributed around the rings. The matrix structures of both the epoxy resins are bisphenol-A type. While this structure contains aromatic rings, the resonant energy is not as high as that of the polyimide. Also, polyimides contain meta-directing groups, which are more stable than ortho- or para-directing groups found in the epoxy resin structure. These effects could explain the enhanced radiation resistance of NP-530, especially compared to G-10CR.

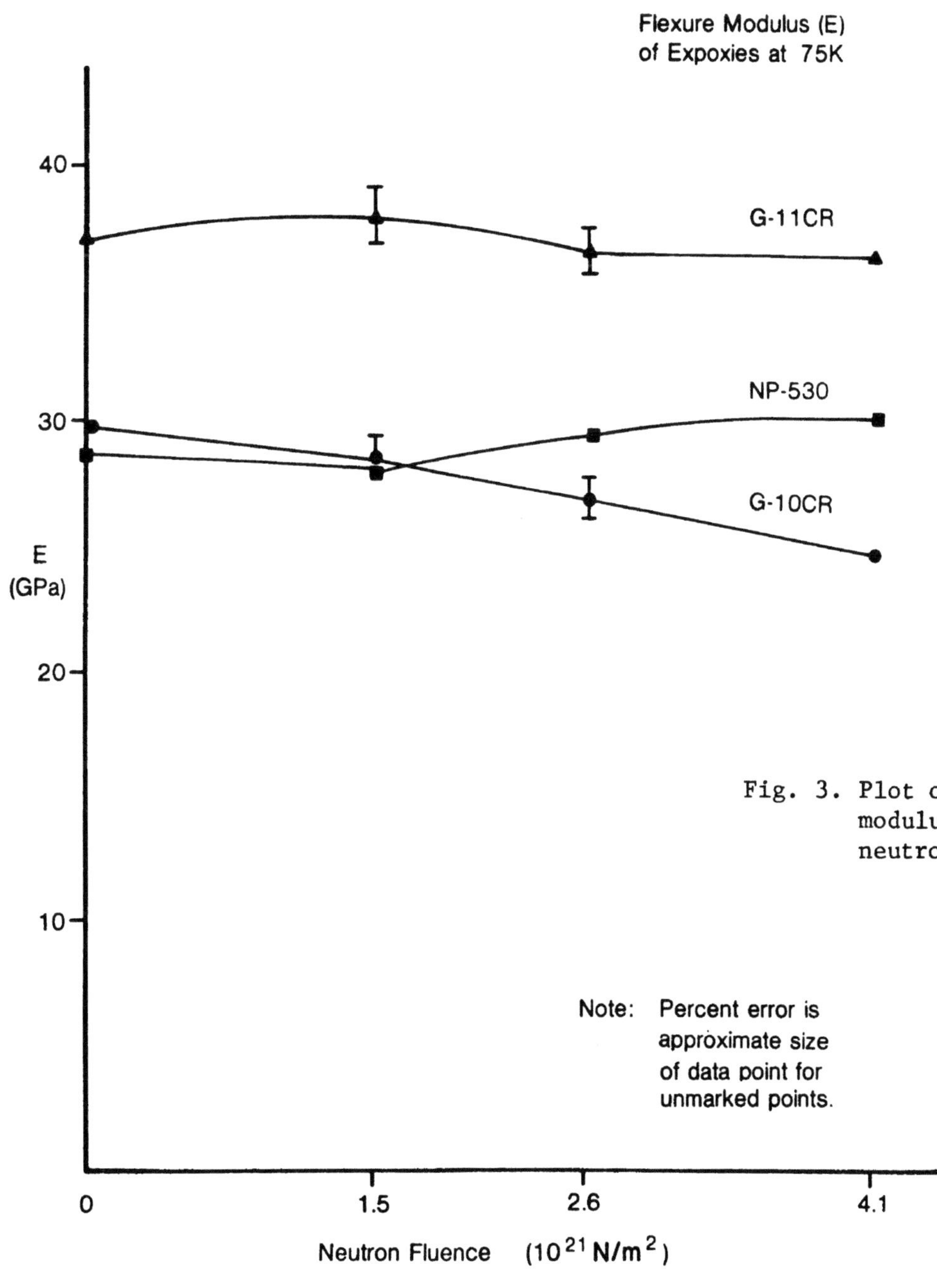

Fig. 3. Plot of flexural modulus versus neutron fluence

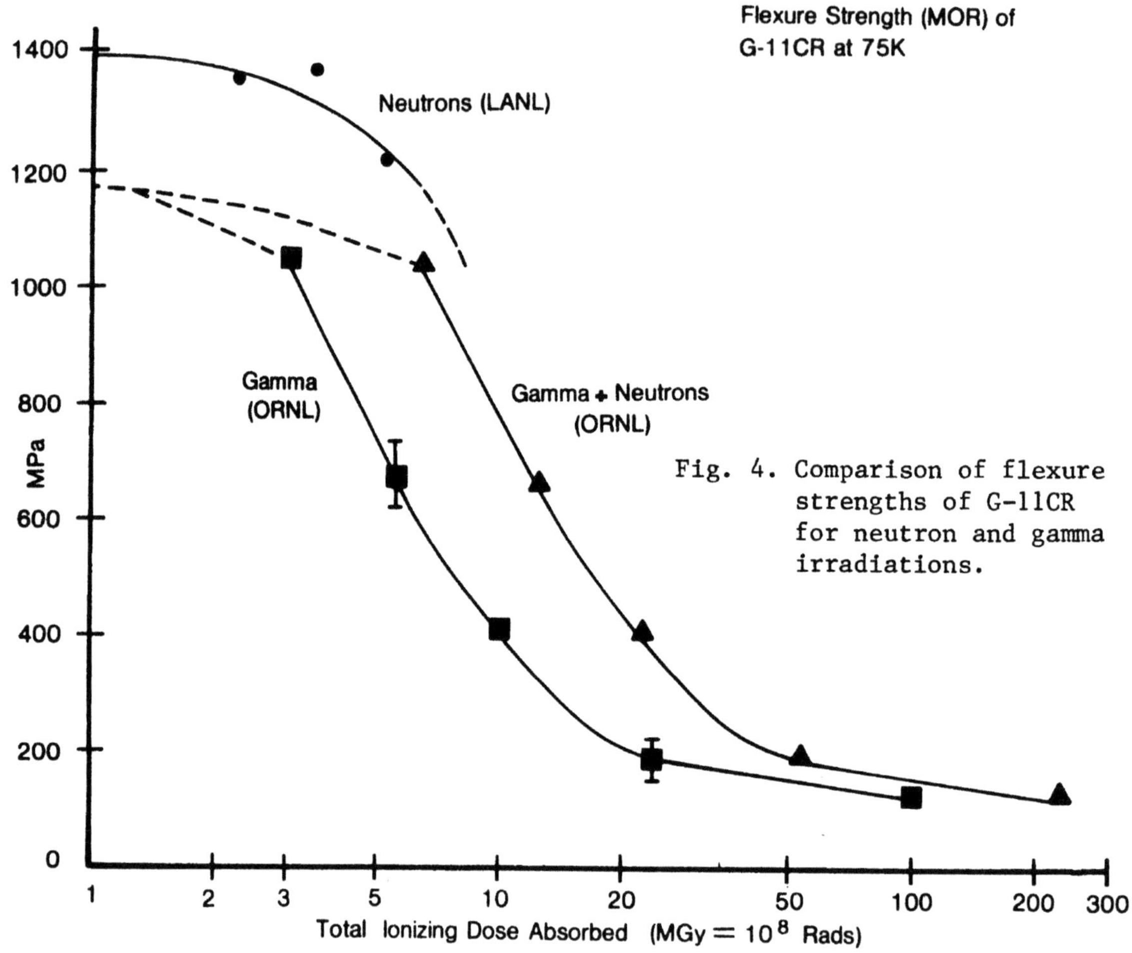

Fig. 4. Comparison of flexure strengths of G-11CR for neutron and gamma irradiations.

A comparison of fast neutron data (this work) with gamma irradiation data for flexure strength of G-11CR is shown in figure 4. As can be seen there is a large discrepency between the strength curves for fast neutron and gamma irradiation. The fiber reinforcement in G-11CR is made of E-glass, which contains 3% boron. The isotope ^{10}B has a large cross section for thermal neutrons resulting in the reaction $^{10}B(n,\alpha)^{7}Li$. If this reaction is taken into account a corrected curve can be generated which more closely agrees with the fast neutron data. A comparison of fast neutron vs. gamma-induced degradation in more damage resistant materials such as NP-530 must await the results of higher dose tests now in progress.

REFERENCES

1. D.C. Phillips, "The Effects of Radiation on Electrical Insulators in Fusion Reactors," AERE-R8923 (1978).
2. "Neutronics Analysis of the Mirror Coil," Fusion Engineering Program, University of Wisconsin, WIS-MARS-82-022, presented at the MARS Preliminary Design Review, Lawrence Livermore Laboratory, Livermore, Ca. (1982).
3. Y. Gohar, "Insulator Materials in Magnetic Fusion Reactors," Proceedings of the Meeting on Electrical Insulators for Fusion Magnets, Germantown, Md., CONF-801237 (1981) pp. 6.1-6.21.
4. B.A. Engholm, "Magnet Shielding and Neutronics Calculations, pp.5.1-5.31, Ibid.
5. D.T. Grubb, J. Mater. Sci., 9:1715 (1974).
6. T. Kato and S. Takamura, "Effect of Low Temperature Reactor Irradiation on Organic Insulators in Superconducting Magnets (I), Cryogenic Engineering 13:243-249 (1978).

7. T. Kato and S. Takamura, " Effect of Low Temperature Reactor Irradiation on Organic Insulators in Superconducting Magnets (II), Cryogenic Engineering 14:178-183 (1979).
8. T. Kato and S. Takamura, "Effect of Low Temperature Reactor Irradiation on Organic Insulators in Superconducting Magnets (III), Cryogenic Engineering 15:173-178 (1980).
9. S. Takamura and T. Kato, Cryogenics 20:441-444 (1980).
10. R.R. Coltman, Jr. and C.E. Klabunde, "Mechanical Strength of Low Temperature-Irradiated Polyimides: A Five to Tenfold Improvement in Dose-Resistance over Epoxies," J. of Nucl. Mat. 103+104:717-722 (1981).
11. S. Takamura and T. Kato, "Mechanical Properties of Organic Insulators for Superconducting Magnets after Low Temperature Irradiation," Presented at the ICMC, Colorado Springs, Aug. 15-19, (1983).
12. R.R. Coltman, Jr. and C.E. Klabunde, "The Strength of G-10CR and G-11CR after Irradiation at 5 K by Gamma Rays," J. of Nucl. Mat. 113: 268-272 (1983).
13. G.F. Hurley and R.R. Coltman, Jr., "Organic Materials for Fusion Reactors," Presented at the Third Topical Meeting on Fusion Reactor Materials, Albuqurque, N.M. Sept. 19-22, (1983).
14. S. Nishijima and T. Okada, "Low Temperature Irradiation Effects on Mechanical Properties of Epoxy Resins used in Superconducting Magnets," Teion Koguka 12:224-229 (1977).
15. R.E. Wyant, Organic Compounds, in: "Effects of Radiation on Materials and Components," J.F. Kercher and R.E. Bowman, eds., Reinhold Publishing Co., New York (1964) pp. 204-276.

A NEW WAY TO CALCULATE THE FRACTURE TOUGHNESS OF MOULDINGS ON IMPACT LOAD IN COMPARISON WITH LINEAR FRACTURE MECHANICS

Hartmut Rest and Georg Menges

Institut für Kunststoffverarbeitung an der
Rhein.-Westf. Technischen Hochschule (IKV)
Pontstraße 49, 5100 Aachen, West Germany

ABSTRACT

The tests for determining fracture data which are described in literature often deal with the methods of linear-elastic fracture mechanics (LEFM). The critical stress intensity factor K_{IC} can only be determined with unusually large material thicknesses and, in addition to that, is strongly dependent on temperature and time. Therefore, great problems are involved in transferring the LEFM-data to thin-wall mouldings. Even more difficulties are encountered in the case of multilayer fiber-reinforced thermoset mouldings, where the LEFM only yields efficient results if cracks propagate between the layers in the resin.

In recent years, the Institut für Kunststoffverarbeitung (IKV) have determined material characteristics in special tensile impact tests. These data - the minimum fracture strain and the minimum volume-specific work to fracture - represent real minima, which are not exceeded even in cases of fastest loading and lowest temperatures. Thus, they are suited for a safe dimensioning against impact. It was proven that the data for mouldings exposed to bending impact can also be safely calculated. Among others a method proposed by Oberbach is used which allows for geometrical influences (direction and location of impact) with the aid of a factor that can be determined during static load or by a FEM-calculation. It is expected that this new way of calculating can also be applied to more complicated mouldings. Relevant tests are being prepared.

PROBLEM

In the past ten years the trend towards substituting other materials with plastics has grown stronger. Primarily in the automobile industry the interest in using thermoplastic mouldings has become more intense, as part of the requirement to save energy by reducing the weight. The trend towards an increased use of plastics is also noticeable in other fields, e.g. pipelines and appliances.

The short glass-fiber-reinforced thermoplastics are suitable in many fields of application, especially where relatively high mechanical damping as well as great ductility and toughness are required.

1. Stress analysis

2. Material selection

3. Determination of permissible material stresses

4. Dimensioning of mouldings

5. Construction of mouldings
Type
Prototype tools and/or production tools

6. Test of mouldings and field tests

7. Production

Fig. 1. Construction appropriate for the material involved according to Goldbach /4/.

It is, however, known that even usually ductile thermoplastics tend to brittle failure in cases of high deformation velocities and low temperatures /1 - 3/.

On the other hand, technically high-grade parts which are exposed to impact are, to an increasing degree, made of plastics.

In order to make sure that the material-specific characteristics are optimally used, the sequence of operations shown in Figure 1 has been followed.

Primarily, the manufacturing of prototypes and their moulds as well as the testing of mouldings prior to start of production are highly expensive in terms of cost and time. A great number of these cost factors could be avoided if more reliable ways of dimensioning were available. This applies particularly for the impact resistance of mouldings for which there exists no practical calculation method. The demand for a reliable method of calculation includes the demand for clearly interpretable and reproducable material data.

STANDARD TESTS WITH IMPACT-EXPOSED MATERIALS

Data and Testing Methods in Accordance with DIN-Standards

Tests for determining the material behaviour in cases of impact loading are most often executed on non-instrumented pendula or falling-weight impact testers, since these machines are relatively low-priced and easy to handle. This is reflected also in the data sheets of the raw material industry. As for high deformation velocities, these sheets usually include only data concerning the fracture toughness and the notched-bar fracture toughness which have been determined in bending impact tests in accordance with DIN 53453 or in conventional Charpy and Izod tests. Sometimes, information is even given on the tensile fracture toughness which has been determined in accordance with DIN 53448 or on the 50 % damage energy during falling-weight impact tests in accordance with DIN 53443 /e.g.5-7/.

Therefore, the aforementioned testing methods are suited solely for obtaining approximate qualitative values and as an aid when deciding on the material suitable for a definite structural component. In any case, the use of these test results as basis of construction is not possible.

Fracture Mechanics

It was expected that the LEFM would be suitable as a method of calculation-of-stability when dimensioning structural components against impact. One objective of fracture mechanics is to find a criterion of judgement as well as to assess whether cracked parts, i.e. predamaged parts, are in danger of fracture /8-10/. In the case of impact loading, the linear-elastic fracture mechanics (LEFM) ought to be a suitable method, since at extreme deformation velocities the marginal case of linear-elastic deformation up to fracture will occur even if thermoplastics are used /11/.

Linear-Elastic Fracture Mechanics

Within the scope of the LEFM, there is a differentiation between the "energy concept", where the fractional behaviour is based on an energy balance for crack propagation, and the "stress-intensity-concept", the laws of which are derived from the elasticity theory. Some aspects of the latter concept will be mentioned here. Detailed descriptions of this subject are to be found in the literature /8-10/.

For measuring the increase in stress at the crack tip, stress intensity factors K are calculated, which are called K_I for simple crack opening through pure tensile stress. The start of instable crack propagation is used as fracture criterion in the case of LEFM. The mathematical condition is:

$$K_I \geq K_{IC} \quad (1)$$

The critical stress intensity, K_{IC}, is called fracture toughness; tests are generally executed in accordance with ASTM E 399-83.

Normally the stress intensity K_I can be calculated as a product of the nominal stress , the square root of crack length "a" and correcting functions. The first difficulties are involved in the

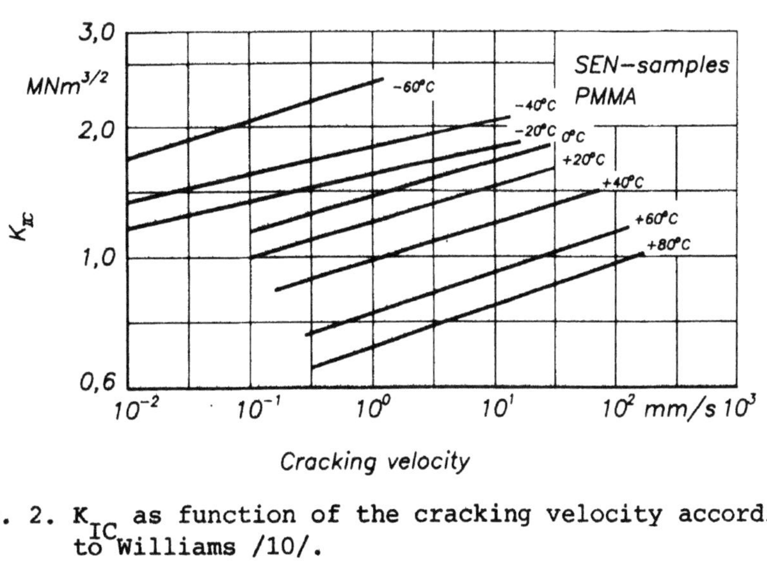

Fig. 2. K_{IC} as function of the cracking velocity according to Williams /10/.

correcting functions, since these need to be newly determined for each individual crack and specimen geometry. In addition to this, K_{IC} values can be determined only with samples of great thickness, since plane strain conditions are taken as the basis for calculation. In /12/, a minimum sample thickness of 12.7 mm is stated for a PE, while the minimum thickness for a PMMA should be approx. 50 mm, as is shown in /13/. A further handicap is the fact that the critical stress intensity factors are strongly dependent on the test temperature and the crack velocity /14/. This phenomenon is shown in Figure 2, giving PMMA as an example.

Even in the case of the lowest tested temperature and the highest rate of crack propagation, the K_{IC} values do not become independent of these parameters. Therefore, it is indispensable to examine these parameter fields as extensively as possible. Once the required test series have been finished, and the determined K_{IC} values are ready to be transferred onto structural components, it is necessary to obtain information on the exact crack geometry which makes possible the calculation of the correcting functions. Moreover, information is required on the stress conditions prevailing in the crack surroundings. These can be either determined approximately, including an accuracy that is difficult to quantify, or by means of time- and cost-consuming calculations, e.g. the finite element method. When all these data and functions are available, statements on the cracking behaviour can be attempted.

This may not, however, necessarily be successful, since contradictions and limits in actual application are found in the literature. According to/15/the LEFM is often better suited for polymers than for metals. On the other hand, /16/ proves that the prerequisites for using the LEFM for thermoplastics are not met. One of the most important requirements is the isotropy of the material. Since this is not the case with all fiber-reinforced thermoplastics, the LEFM cannot be applicable to this important material group. As for fiber-reinforced thermosets, the LEFM can only be applied in such areas, which consist exclusively of resin between the fiber layers, so that crack advance is not affected by fibers.

In fact, hardly any technical publication can be found as to fracture mechanics tests of fiber-reinforced thermoplastics. One article /17/ deals with K_I-measurements of carbon fiber-reinforced polycarbonate. The sample thickness of approx. 3.2 mm was, however, so unfavourably chosen that it was impossible to determine the K_{IC}-values.

Moulding Construction and LEFM

The question is, whether the LEFM data could be of use when dimensioning new mouldings.

The dimensioning of mouldings is based on ideal parts. Thus, possibly available K_{IC}-values cannot be used without further checking, since - aside from the isotropy condition - the basis of application, namely the crack, does not exist. If the LEFM-method is used throughout, a length being characteristic of the material or the moulding surface would have to be assumed for a fictional crack length. The question is, which length.

In the case of fiber-reinforced thermoplastics, the fiber geometry could be used for defining a crack length, whereas a definition of the surface roughness could be applied with non-reinforced material.

Apparently, no investigation into this subject has been undertaken to date, although fracture mechanics are much older than plastics. Therefore, probability of success is at least doubtful.

Hence, the need for a reliable way of dimensioning is still existing.

THE NEW METHOD OF DIMENSIONING

In the course of developing this new method, it has been necessary to establish, first of all, safe and definite material data. Easily interpretable tests were to be executed in the shortest possible time. The calculation of the moulding failure was to be effected with the minimum investment in equipment and time necessary.

The following chapters show how these requirements were met.

Data and Method of Calculation

In recent years, the Institut für Kunststoffverarbeitung have used hydraulic testers for determining the impact behaviour of thermoplastic materials. Two different tensile tests have been carried out on dumb-bell samples:

- creep tests over 10^3 seconds
  (loading time: 5 to 10 milliseconds)
- high-velocity tearing tests
  (time to breaking: about 10 milliseconds)

While recent creep tests were primarily executed at ambient temperature, the high velocity tearing tests were carried out at temperatures varying from ambient to -180 °C. A detailed description of the tests is given in /11/.

Minimum Fracture Strain

In the course of investigation, the fracture strain proved to be highly dependent on temperature and time; first tested was PMMA of high molecular weight /2/, and later materials such as PVC-rigid, POM and glass-fiber-reinforced PA6 /18/.

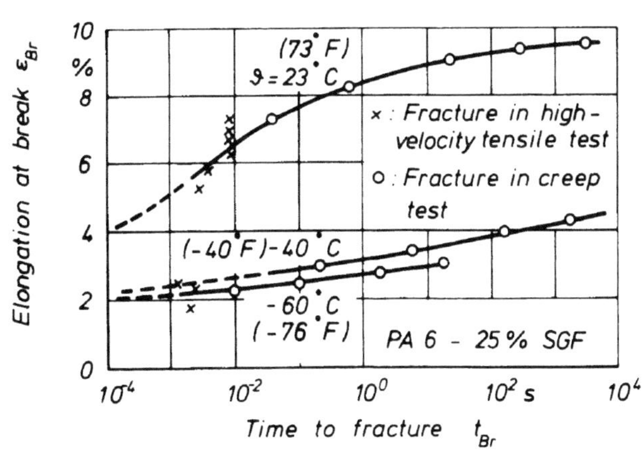

Fig. 3. Elongation at break versus time to fracture in creep test under static load (PA6-25%SGF)

As is shown in Figure 3, the fracture strain decreases and approaches a marginally acceptable value within ranges of time and temperature depending on the material. This lowest experimental fracture strain value has been called minimum fracture strain. This trend was recently proven also in the cases of PC, glass-fiber-reinforced PC and ABS /3/.

As the value of the minimum fracture strain does not decrease even in cases of highest deformation velocities and lowest temperatures, the minimum fracture strain is a real, safe marginal value, which can be used for transferring the principle of strain-oriented dimensioning /19/ onto the impact loading case.

For the purpose of determining the minimum fracture strain, nothing is required but the execution of high-velocity tearing tests or bending impact tests at a temperature near that of liquid nitrogen.

Practical cases like notches etc., however, cannot be calculated with the aid of a marginal strain value. Therefore, another associated parameter must be found, e.g. the critical energy.

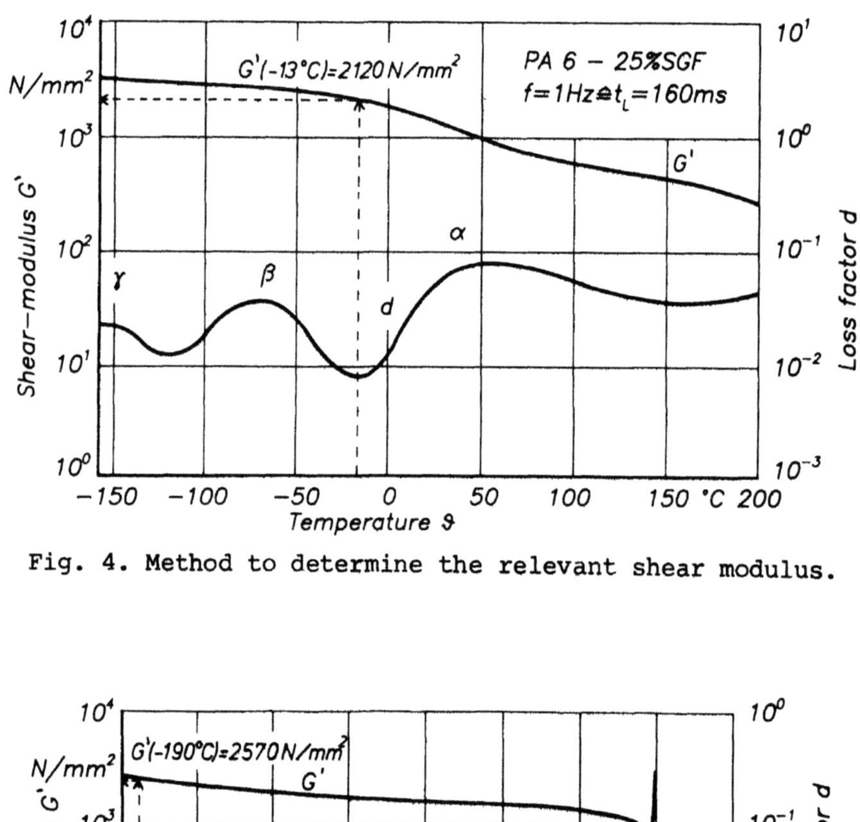

Fig. 4. Method to determine the relevant shear modulus.

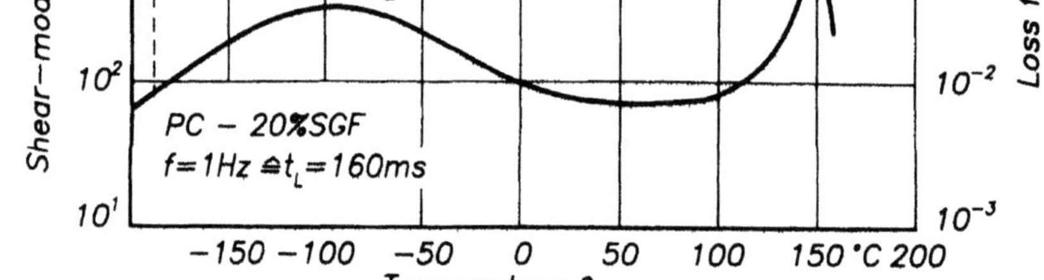

Fig. 5. Method to determine the relevant shear modulus.

Young's Modulus

In addition to the minimum fracture strain, Young's modulus is required as a second parameter. With respect to the estimating character of this calculation it seems to be appropriate to determine the Young's modulus from the shear modulus diagrams that are available for all plastics.

Proceeding from the assumption that time and temperature exert an influence in the same direction on the mechanical characteristics of thermoplastics /e.g.20-23/, the generally available shear-modulus values were converted into Young's moduli with the aid of the method described by Boden /11/. The calculation was based on the formula given by /22/ for the linear-elastic region:

$$E_0 = 2\ G'\ (1 + \nu) \qquad (2)$$

with $\nu = 0.37$.

This is legitimate, because the material has a completely linear-elastic behaviour in the critical case of brittle fracture due to impact.

Table 1 contains a comparison between the shear moduli and the Young's moduli deduced for various materials.

Torsion pendulum tests yielded the G'-values, which were read off when the damping maximum was exceeded in the temperature range of application, or, if a damping maximum did not occur, the values were read behind the first maximum registered at lower temperatures. The procedure is schematically shown in Figure 4 for semi-crystalline PA6 containing 25 per cent of short glass fibers (SGF) and in Figure 5 for glass-fiber-reinforced PC.

Additionally, the reference temperatures for high-velocity tearing tests with a time to fracture of about 10 ms that had been calculated with the aid of the time-temperature shifting factors given by /11/, and the Young's moduli of the stress-strain-diagrams, having been determined by experiment on the first occurence of $\varepsilon_{fracture} = \varepsilon_{min}$, i.e. the quotients $\sigma_{min}/\varepsilon_{min}$ were listed. These comparative coefficients of temperature and velocity are required, because the shear-modulus-diagrams are available only as a function of temperature. The comparison between calculated and measured

Table 1. Comparison of Young's moduli calculated from shear moduli to experimentally derived E_{min}-values

Material	TORSION PENDULUM TEST ($t = 1/2\pi f$ = 160 ms) Temp. °C	Shear modulus G' $N\ mm^{-2}$	shifting factor acc. to /11/ °C/Decade	equivalent temp. °C	Calculated Young's modulus E_0 $N\ mm^{-2}$	HIGH-VELOCITY TEARING TEST ($t \approx 10$ ms) Test temp. °C	measured Young's modulus E_{min} $N\ mm^{-2}$
PMMA 233	-10	1975	21,4	+ 16	5410	23	6136
PA6-25% SGF	-13	2120	10,1	- 1	5810	-60	9720
PVC-rigid	-145	2050	15,5	-126	5610	-120	6250
PC	-190	1645	11,6	-176	4510	-180	4550
PC-20% SGF	-190	2565	11,6	-176	7030	-170	7200

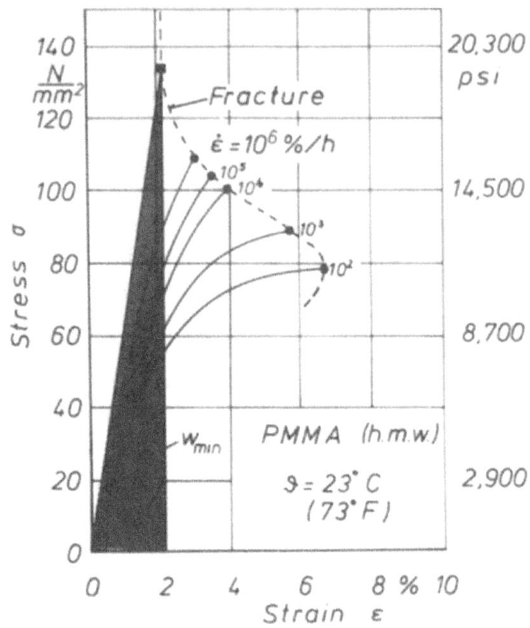

Fig. 6. Limits of stress and strain of PMMA on impact.

Young's moduli only for PA6-25 % SGF reveals a deviation of more than 50 per cent related to E_o. An explanation for this phenomenon has not yet been found. It may be caused by a disadvantageous influence of fiber-orientation and/or conditioning, i.e. the moisture content of the material.

The other values show a good to very good conformity. In any case, even for PA6 with SGF, the Young's moduli deduced from the shear moduli are below the measured values, i.e. on the safe side. This fact is also important with respect to the applicability of this way of dimensioning.

Minimum Volume-Specific Work-to-Fracture

Figure 6 shows, by the example of PMMA, the results of tensile tests at various strain velocities. It can be seen that the decreasing fracture strain is accompanied by an increasing fracture stress. (This is by the way a proof of the inapplicability of a stress parameter in the course of dimensioning).

With varying deformation velocities, the area under the stress-strain curve shows clearly that the volume-specific work-to-fracture (w) first declines with increasing velocity and finally arrives at a minimum value w_{min} when reaching the minimum fracture strain. The relationship is shown in Figure 7.

Since the minimum fracture strain value does not decrease even under extreme conditions, the volume-specific work-to-fracture must increase, because of an increasing fracture stress.

For impact loading case, a theory can be derived from the above relationship and from Chapter 3.1.1: A moulding will not fail unless the minimum fracture strain or the minimum volume-specific work-to-fracture is exceeded in highest-stressed part of the moulding.

If the parameter w_{min} is used for dimensioning, the most un-

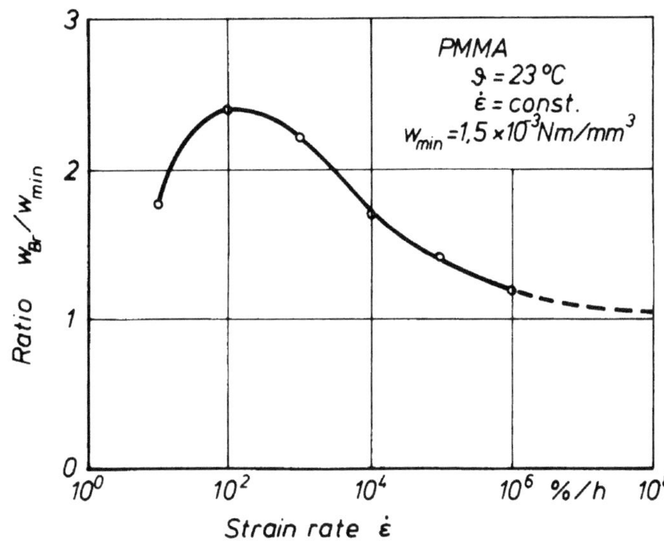

Fig. 7. Normed volume-specific work to fracture with $\dot{\varepsilon} \geq 10$ %/h.

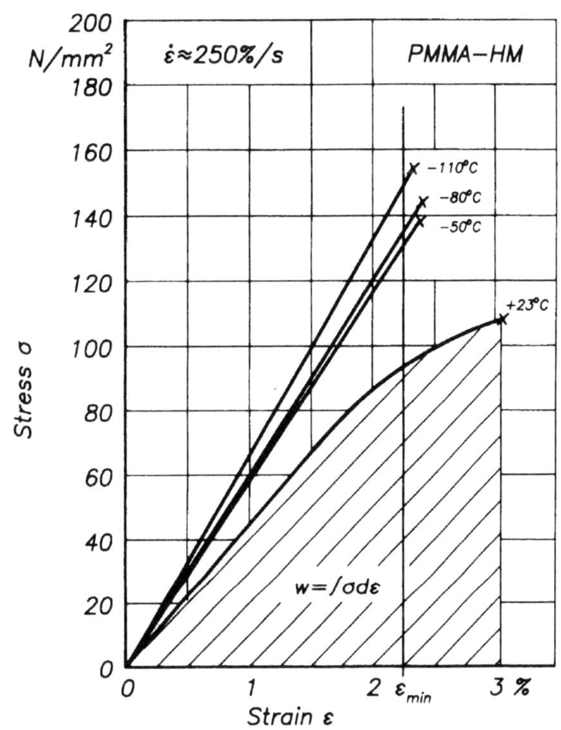

Fig. 8. Stress-strain behaviour of PMMA at varying temperatures.

favourable case is taken into account, thus always yielding safe results.

Since the experimental determination of w_{min} would require a great number of test series, a simpler method was sought.

The volume-specific work-to-fracture w is defined as

$$w = \int_0^{\varepsilon} \sigma \; d\varepsilon' \qquad (3)$$

Hence, it represents the area below the stress-strain-diagram. If the relationship is linear, equation (3) may be simplified to:

Table 2. Comparison between experimentally determined and calculated volume-specific work to fracture of different thermoplastics

Material	Test ε_{min} %	$w_{min}=\frac{1}{2}\sigma_{min}\cdot\varepsilon_{min}$ Nmm/mm³	Calculation $w_{Br}=\frac{1}{2}E_0\cdot\varepsilon_{min}^2$ Nmm/mm³	w_{Br}/w_{min}
PMMA	2,2	1,49	1,31	0,88
PA6-25% SGF	1,8	1,58	1,03	0,65
PVC-rigid	2,0	1,25	1,12	0,90
PC	5,0	5,75	5,64	0,98
PC-20% SGF	3,0	3,23	3,16	0,98

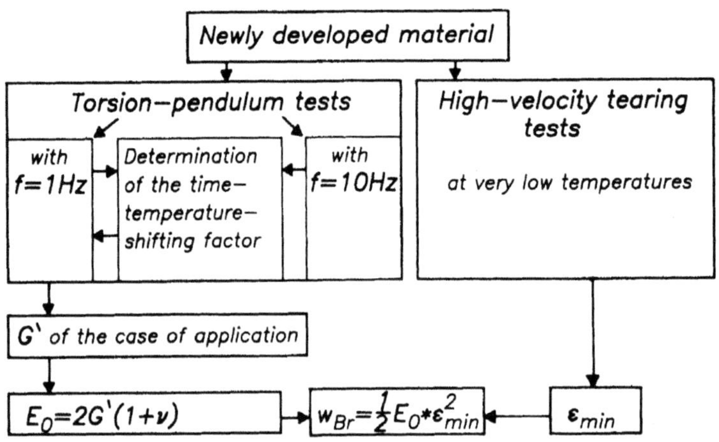

Fig. 9. Pattern for determining impact characteristics E_0, w_{Br}, ε_{min}

$$w=\frac{1}{2}\sigma\cdot\varepsilon \tag{4}$$

This simplification is legitimate in this case, because the drop of the fracture strain is coupled with a corresponding embrittlement of the material, which can be seen a.o. from the growing linearization of the stress-strain-behaviour.

If equation (4) is combined with Hooke's Law and the stress function is eliminated, the following equation is obtained:

$$w=\frac{1}{2}E_0\cdot\varepsilon^2 \tag{5}$$

Compared with the E-modulus, the minimum fracture strain can be determined with greater reliability. It is entered as a squared number into equation (5) so that an exact estimation of w_{min} should be possible within the given boundaries.

Table 2 contains a comparison of the experimentally determined w_{min}-values and the data calculated in accordance with equation (5).

The comparison reveals that the work-to-fracture values w_{Br}, which were determined using the method described above, are smaller than the w_{min}-values. Except for the PA6 -25 % SGF, the deviations do not exceed 12 per cent. This excellent conformity is remarkable, especially in view of the small number of tests required for determining the w_{Br}-values.

Practical Approach

Should the practician be in need of a quick survey on the impact loading characteristics of a new material as well as of parameters for safe dimensioning, the pattern shown in Figure 9 will be of use.

In addition, the time-temperature shifting factor is required to enable integration of the assumed time and temperature conditons into the torsion-pendulum tests. Thus, it serves the purpose of fixing the shear modulus relative to the practical case. This factor can be defined by comparing the damping curves which are obtained through torsion-pendulum tests at two different frequencies. For calculating Young's modulus in the impact loading case considered here, the Poisson's ratio should be 0.37 as shown in /11/.

Calculation of the Earliest-Possible Failure of Mouldings under Impact Loading

Now that methods of determining the parameters have been discussed, their use for impact dimensioning shall be outlined.

A great problem involved in the usual impact tests is that the measured values cannot be easily transferred from the specimen to the moulding to be computed. A method of calculating the volume-specific work-to-fracture of the highest-stressed volume element has been presented in /24/. This does, however, require that the deformation characteristics can be calculated, either by way of transferring Hooke's Law onto simple geometries, or, in the cases of complex mouldings, by using the finite-element method. If necessary, it is also possible to dissect the mouldings into segments which can be computed individually. The geometric conditions of the load case to be calculated, as well as of the moulding, are included in /24/ in the form of a factor "K". The total work-to-fracture W is connected with the volume-specific value w through the K-factor, having the dimension of a volume:

$$W = K \cdot w \qquad (6)$$

In addition to the occuring stresses, the strains resulting from the assumed load case and the volume-specific work-to-fracture are to be determined in the course of calculation and to be compared with the minimum acceptable values. Should the calculated values be greater than the minima, an improvement can be achieved by changing the geometry or selecting another material, while an optimization would be possible through repeated calculations. In addition to the above, structural boundaries have to be taken into account, e.g. maximum deflections and existing notches.

Application of Method and Parameters to Plates Exposed to Bending Impacts

The calculation method described above was first scrutinized using the simplest structural component, i.e. the plate.

The check was executed by comparing experimental data and calculated values. The following experimental parameters were used:

Material:	PA6 -25 % SGF
Temperature:	-60 °C

Mounting distance:	70 mm
Sample length:	100 mm
Sample width:	60 mm
Sample thickness:	4,2 mm
Breaking time:	about 10 ms
Loading place:	central, linearly across the full sample

Figure 10 shows the formula and data on which the calculation was based. The ratio between the calculated fractional values of deflection, force and work-to-fracture and the experimental values will always yield a factor above one. Thus, one is always on the safe side. This also applies to the materials PMMA and PVC-rigid which have been tested under the same conditions (Table 3). The volume-specific work-to-fracture factor is greater than four. The reasons for this relatively great deviation are presently being examined.

In principle, this result is comparable with the findings of /18/ that during the bending impact test crazes occured in the boundary fiber of the PMMA exposed to tensile strain, while completely brittle failure was observed during the tension test, under equal conditions of time and temperature. The occurence of the crazes corresponds to a higher work-to-fracture in the case of a bending load. In this context it has to be tested whether an internal supporting effect /25/, or a non-linear distribution of stress and strain over the sample thickness or in the direction of the longitudinal axis of the sample produce an unexpectedly great experimental load bearing capability. The latter cannot be precluded in view of the results stated in /26/.

It is assumed that the fracture strain during the flexural test is not higher than during the tensile test under equal boundary conditions. This assumption is based on the results gained with strain gauges which were tested on the same tester at ambient temperature. Bending impact tests at extremely low temperatures, including strain control, are in the preparatory stage.

FORECAST

It has been proven that the method presented here allows mouldings to be dimensioned safely, even if they are made of fiber-reinforced thermoplastics. Further scientific work is planned and preparations are being made to determine the degree of reliability and the reason for the reliability of this new method of calculation.

It should also be possible to calculate mouldings with inhomogeneous wall structure, such as multilayer laminates or sandwich

Table 3: Comparison measurement/calculation for bending impact exposed plates

Material	f_{Br} / Calcul.	mm Exp.	F_{Br} / Calcul.	N Exp.	W_{Br} / Calcul.	Nmm Exp.
PA6-25 % SGF	3,5	9,2	1155	1870	2020	8600
PMMA	6,0	10,2	612	852	1836	4473
PVC-rigid	4,1	15,7	1026	1913	2093	15000

ϑ = -60 °C t_{Br} = 10 ms   Three-point-bending

$t_{Br}=10ms$ | $l=70mm$

$\vartheta=-60°C$ | $b=60mm$

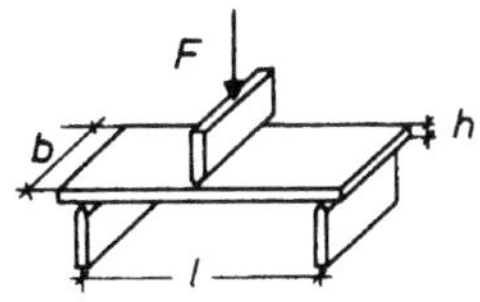

PA 6–25%SGF | $h=4,2mm$

from high–velocity tearing tests : $\varepsilon_{min}=1,8\%$

from torsion–pendulum tests : $E_0=6360\ \frac{N}{mm^2}$

Calculation		Test /11/
$f_{Br}=\frac{\varepsilon_{min}*l^2}{6h}$	$=3,5mm$	$f_{Br}=9,2mm$
$F_{Br}=\frac{4\varepsilon_{min}*E_0*W_b}{l}$	$=1155N$	$F_{Br}=1870N$
$W_{Br}=\frac{1}{2}F_{Br}*f_{Br}$	$=2020Nmm$	$W_{Br}=8600Nmm$
$w_{Br}=\frac{1}{2}E_0*\varepsilon_{min}^2$ $=1,03\frac{Nmm}{mm^3}$	$=\frac{W_{Br}}{K}$	$w_{Br}=4,39\frac{Nmm}{mm^3}$

Fig. 10. Comparison measurement/calculation of a plate exposed to bending impact.

elements. In these cases the layers would have to be examined consecutively, taking the weakest layer for determining the criteria of failure.

Thus, it would be possible to dimension structural components of all polymer materials as well as their combinations with the aid of a single concept.

ACKNOWLEDGEMENTS

We express our thanks for the important financial assistance during the major part of this work to the 'Arbeitsgemeinschaft Industrieller Forschungsvereinigungen' (AIF) and to the 'Deutsche Forschungsgemeinschaft' (DFG). For the gratuitous donation of test material we owe our thanks to BASF, Ludwigshafen, Bayer AG, Leverkusen, Chemische Werke Hüls AG, Marl, and Röhm GmbH, Darmstadt.

REFERENCES

1. H. E. Boden, "Bruch- und Verformungsverhalten von Thermoplasten beim Kriechen im Zeitbereich von 10^{-3} bis 10^{3} s", Final report, "DFG/Me 272/144", IKV, Aachen, 1980
2. B. Kleinemeier, H. E. Boden, "Das Spannungs-Dehnungsverhalten von Thermoplasten bei Stoßbeanspruchung", Final report "AIF 4094", IKV, Aachen, 1980
3. A. Paulus, unpublished work, IKV, Aachen, 1984
4. H. Goldbach, "Werkstoffgerechtes Konstruieren", Anwendungstechnische Information 380/83, Sparte KL, Bayer AG, Leverkusen, 1983
5. B. Carlowitz, "Das Verhalten von Polystyrol unter biaxialer Schlagbeanspruchung", Plastverarbeiter 30 (1979) 7, p. 377-381
6. H. P. Gilfrich et al, "Untersuchung des Stoßverhaltens von Kunststoffen mit dem Fallbolzenversuch und dem Durchstoßversuch mit

elektronischer Meßwerterfassung", Kunststoffe 63 (1973) 5, p. 327-333
7. J. J. Mooij, "Instrumented flat-headed falling-dart test", Polymer Testing 2 (1981), S. 69-83
8. H. Blumenauer, G. Pusch, Technische Bruchmechanik, VEB Deutscher Verlag für Grundstoffindustrie, Leipzig 1981
9. W. Dander, "Einführung in die Grundlagen der Bruchmechanik", Z. Werkstofftech. 13 (1982), pp. 69-76, pp. 85-95, pp. 385-388
10. F. Kerkhof, "Anwendung der Bruchmechanik auf Hochpolymere", Kolloid-Z. u. Z. für Polymere 251 (1973) 8, pp. 545-553
11. H. E. Boden, "Das mechanische Verhalten von Thermoplasten bei stoßartiger Belastung", thesis, IKV, RWTH Aachen, 1983
12. J. F. Mandell, D. R. Roberts, F. J. Mc Garry, "Plane strain fracture toughness of polyethylene pipe, Polymer Eng. and Science, 23 (1983) 7, pp. 404-411
13. U. Meier, A. Rösli, "Linear-elastische Bruchmechanik an polymeren Werkstoffen", Material und Technik 6 (1978) 4, pp. 148-158
14. J. G. Williams, "Application of linear fracture mechanics", in "Advances in Polymer Science" 27, pp. 67-121, Springer Verlag, Berlin 1978
15. J.G. Williams, "Application of linear fracture mechanics", Polymer Eng. and Science 17 (1977) 3, pp. 144-149
16. H. J. Schwalbe, "Promotionsvortrag", IKV, RWTH Aachen, 1981
17. W. E. Haskell III., S.P. Petrie, R.W. Lewis, "Fracture toughness of short-fibre reinforced thermoplastic", Polymer Eng. and Science 23 (1983) 14, pp. 771-775
18. H. E. Boden, "Stoßverhalten von Thermoplasten unter Berücksichtigung von Glasfaserverstärkung, Kerben und tiefen Temperaturen", Final report "AIF 4707", IKV, Aachen, 1982
19. G. Menges, R. Taprogge, "Denken in Verformungen erleichtert das Dimensionieren von Kunststoffen", VDI-Zeitschrift 118 (1970), pp. 341-346 u. pp. 627-629
20. D. W. van Krevelen, Properties of Polymers, Elsevier, Amsterdam, 1976
21. G. Menges, Werkstoffkunde der Kunststoffe, Carl Hanser Verlag, München, 1979
22. G. Schreyer, Konstruieren mit Kunststoffen, Band I und II, Carl Hanser Verlag, München, 1972
23. H. H. Kausch, Polymer Fracture, Springer Verlag Berlin, 1978
24. K. Oberbach, "Zur Aussagekraft von Stoßversuchen an Probekörpern für das Zähigkeitsverhalten von Plastformteilen", Plaste u. Kautschuk 29 (1982) 3, pp. 178-182
25. E. Siebel, "Neue Wege der Festigkeitsrechnung", Z. des VDI, 90 (1948) 5, pp. 135-139
26. F. Eversmann, Studienarbeit, IKV, 1984, Archiv-Nr. S 8433, Supervisor: H. Rest

CRAZE FORMATION AND CRYOGENIC TOUGHNESS

W. Döll, L. Könczöl, and M.G. Schinker

Fraunhofer-Institut für Werkstoffmechanik

Wöhlerstr. 11, D - 7800 Freiburg, FRG

Abstract

Crazing is the predominant plastic deformation mechanism in thermoplastics preceeding brittle fracture. The energies needed to form and to stretch crazes up to their break also mainly determine the fracture energy. Using the fracture mechanics approach together with the Dugdale type of plastic zone the single craze at a crack tip can be well characterized as may be seen from experimental investigations using the microscopic optical interference technique. At low temperatures the good deformation behaviour of PC is associated with the crazing mechanism.

Introductionary remarks on crazing

Thermoplastic materials exhibit an interesting mechanical behaviour at large irreversible deformations in forming shear zones and/or craze zones. Both processes are yield phenomena, however, the shear zones are induced - as is already indicated by the word - by shear stresses, while craze zones are induced by tensile normal stresses. Both of these processes are to be seen as precursors of fracture and - furthermore - determine the main part of the fracture energy. While fracture after shear yielding normally can be characterized as fracture consuming high energies, fracture at crazing is connected with lower energies and is characterized as brittle. Since brittle fracture and hence crazing is favoured by high loading rates and/or low temperatures we will focus our interest in this paper on the phenomenon of crazing and its relation to the fracture process*).

To demonstrate the stress situation which favours or prevents crazing Fig. 1 shows an example of a PMMA specimen with a circular hole which has been loaded by uniaxial tension in the vertical direction indicated by arrows. In this loading situation there is a pressure region above and be-

*) For a deeper insight into the crazing phenomenon the detailed reviews in this matter in the book "Crazing in Polymers" [1] with many further literature cited is recommended.

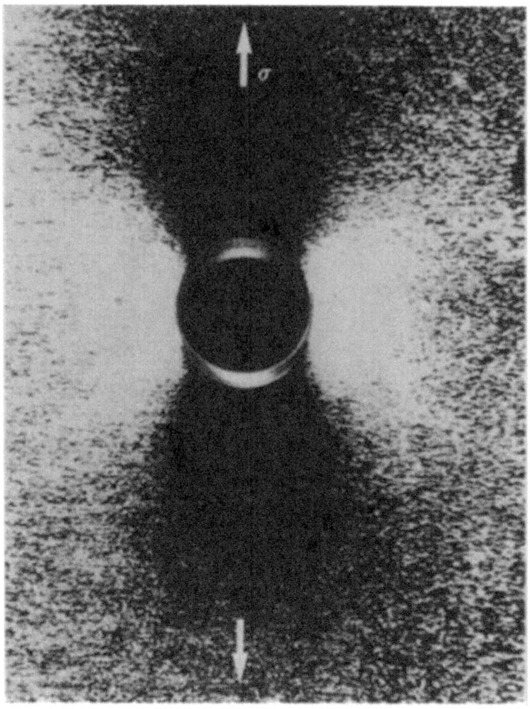

Fig. 1: Crazes around a hole in a PMMA specimen under uniaxial external tensile load.

low the hole while on both sides regions with tensile stresses exist having a maximum on the side rims of the hole whose magnitude is the threefold of the external stress.

Taking the photograph of the loaded specimen in reflected light the crazes are made visible in the bright areas in which tensile stresses act. In the dark areas where thus no crazes arise pressure stresses dominate. It can clearly be seen that the craze density is highest where high tensile stresses act such as at the side rime regions of the hole.

Crazes can be described as long, thin zones ranging from some µm to some cm in length while the thickness is in the order of µm. It has been stated that craze zones are perpendicular to the major principal tensile stress and grow into the direction of the minor principal tensile stress. Although looking in their geometrical form similar to cracks craze differ essentially from cracks since they contain stretched material (fibrils, as may be seen from Fig. 2 [2]) oriented in the direction of the major principal tensile stress. The fibrils are surrounded by voids thus leading to a lower density and hence a lower refractive index of the craze zone than of the bulk material. The density of the craze zone will decrease with draw ratio. A typical void content in the craze zone may be in the order of 50 to 80 %, depending on the material and on the conditions of craze growth.

If a craze is broken the fibrillar structure of the craze should also be reflected by its remmants on the fracture surfaces. Under certain circumstances it has been achieved to document these structures. Figs. 3 and 4 show different types of single fibrils of the craze material broken under fatigue loading during discontinuous crack growth. While Fig. 3 shows short, closely crowded fibrils from the region of the craze tip Fig. 4 shows long widely spread fibrils from a region of larger craze thickness. Figs. 3 and 4 also demonstrate that during fracture - that is the propagation of cracks - crazing occurs at the crack tip. By fracture mechanics a tool has been established to calculate fracture energies to describe crazing at the crack tip analytically.

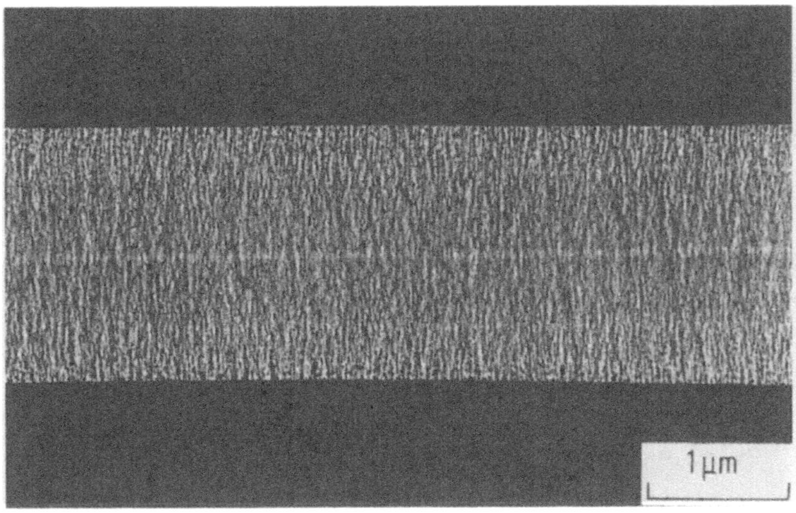

Fig. 2: TEM micrograph of a PS craze fibril micro-structure (M.B. Yaffe and E.J. Kramer, J. Mater. Sci. 16 (1981) 2139, Chapman and Hall, Ltd.)

<u>Analytical description of crack tip crazing</u>

In this paper only crack openings in response to tensile stresses normal to the crack plane will be considered in detail. This is known as mode-I loading. Under these conditions the normal stresses σ_{yy} in the crack plane in front of the crack tip are given for an ideal elastic material by:

$$\sigma_{yy} = \frac{K_I}{\sqrt{2\pi r}} \qquad (1)$$

Fig. 3

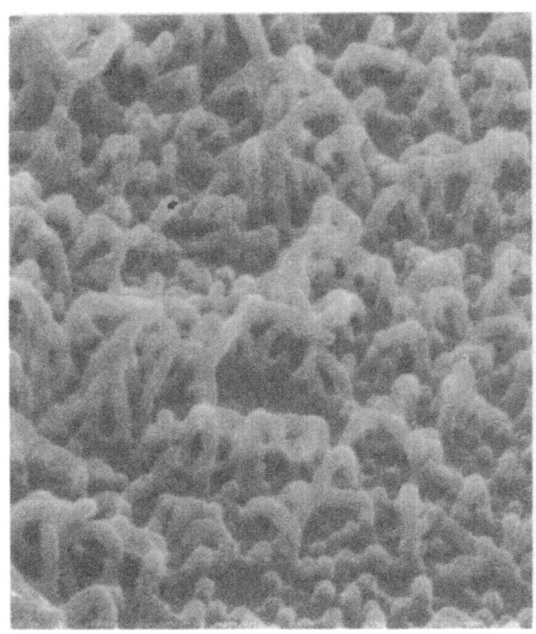

Fig. 4

Figs. 3 + 4: SEM micrographs of the fibrillar microstructure in fatigue fracture surfaces in PVC at different positions of the former craze zone.

The dependence of σ_{yy} on the distance r from the crack tip is qualitatively shown in Fig. 5, which also indicates that the stresses have a singularity at the crack tip and that K_I determines the stress intensity in the crack tip region. The stress intensity factor K_I describes the combined effect of the external stress σ and the crack length a:

$$K_I = \sigma \cdot \sqrt{a} \cdot Y \tag{2}$$

where Y is a correction factor which takes into account the particular configuration of specimen and crack geometry.

The opening 2v(r) of a crack in a linear elastic material is given by

$$2v(r) = \frac{8\,K_I}{E^*} \cdot \sqrt{\frac{r}{2\pi}} \tag{3}$$

where E* is the reduced modulus ($E^* = E$ in plane stress and $E^* = E/(1-\nu^2)$ in plane strain, ν = Poisson's ratio).

From Eq. (3) and also from Fig. 5 it can be seen that near its tip the crack opens parabolically; the magnitude of the opening 2v(r) at a fixed distance is influenced by the elastic properties of the material and governed by the stress intensity factor K_I.

The behaviour of propagating cracks is often described in terms of the elastic strain energy required to create unit area of fracture surface: the strain energy release rate G_I. Between stress intensity factor K_I and strain energy release rate G_I the following relationship exists:

$$K_I^2 = G_I \cdot E^* \tag{4}$$

It is now well documented that a good description of the shape and size of this yielded zone at the crack tip can be provided by the plastic zone size model proposed by Dugdale [3]. In this model which is similar to the cohesive force model of Barenblatt [4] and others the length s of the plastic zones is determined by the condition that there should be no stress singularity at the tip of the crack.

For small scale yielding (i.e. $\sigma \ll \sigma_c$ = craze stress) this can be expressed in fracture mechanics terms as:

$$S = \frac{\pi}{8} \cdot \frac{K_I^2}{\sigma_c^2} \tag{5}$$

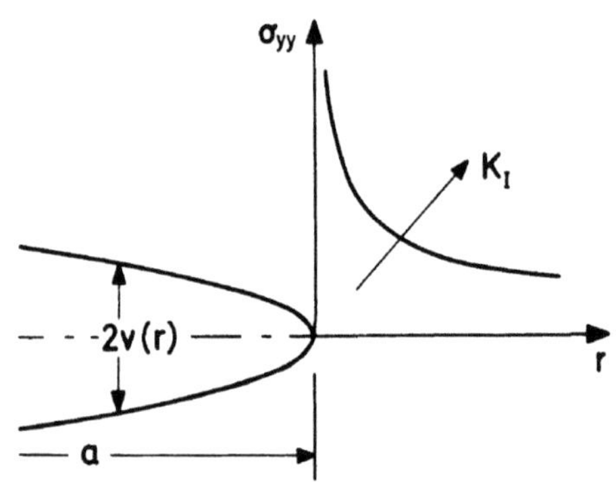

Fig. 5: Crack opening and tensile stress distribution near the crack tip in an ideal elastic material

An analytical expression for the displacements 2v of the elastic-plastic boundaries in front of and behind the crack tip was derived by Goodier and Field [5].

A simpler expression for the displacements of the plastic zone boundaries alone has been derived by Rice [6]. It has the form

$$2v(x_1) = \frac{8\,\sigma_c \cdot s}{\pi \cdot E^*} \cdot \left\{ \xi - \frac{x_1}{2s} \ln\left(\frac{1+\xi}{1-\xi}\right)\right\} \qquad (6)$$

where $\xi = (1-\frac{x_1}{s})^{1/2}$ with x_1 = coordinate parameter, crack tip at $x_1 = 0$.

In Fig. 6 the dependence of the displacements on the distance from the crack tip is shown using a normalized scale and assuming $\sigma = 0.1\,\sigma_c$. At a short distance behind the crack tip the elastic-plastic boundary has an almost parabolic form, whilst directly at the crack tip - in contrast to the ideal elastic solution (Eq.(3)) - there is a certain displacement which is also the maximum width of the plastic zone. Here, and in the following text, the width of the plastic zone at the crack tip will be denoted by 2v without any coordinate parameter. In fracture mechanics terms it can be expressed as

$$2v = K_I^2/(\sigma_c \cdot E^*) \qquad (7)$$

Hence the displacement 2v at the crack tip is governed not only by the material's properties (E^*, σ_c) but also, and more severely, by the stress intensity factor K_I.

Using the relationship between stress intensity factor K_I and strain energy release rate G_I Equation (6) may be rearranged in the form:

$$G_I = 2v \cdot \sigma_c \qquad (8)$$

which is of special importance for the interpretation of fracture in polymers on molecular basis.

A note should be added concerning the two parameters which chiefly characterize the shape of the plastic zone: length s and maximum width 2v.

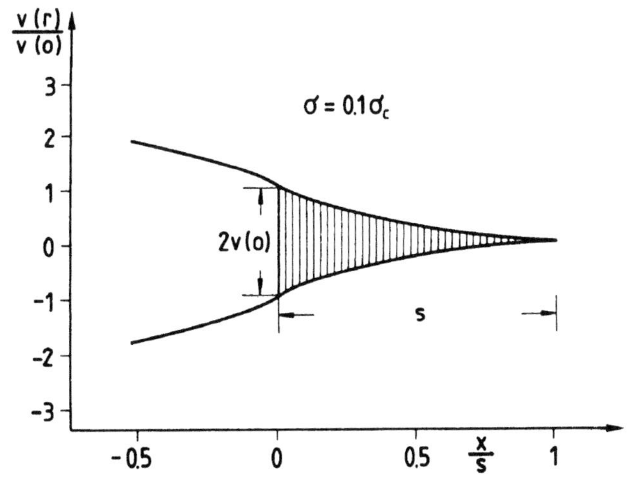

Fig. 6: Elastic-plastic boundary as calculated from the Dugdale-model

Whilst both depend strongly on K_I (compare Eq. (4) and (6)) the relationship of maximum width to length does not:

$$\frac{2v}{s} = \frac{8}{\pi} \cdot \frac{\sigma_c}{E^*} \tag{9}$$

It is only dependent on the material's properties.

Experimental determination of crack tip craze zones

The classical methods of light optical interference are well known in the determination of small dimensions which are of the order of the wave length of light. A new scientific field has been opened up by their application to investigations of the craze behaviour at crack tips in transparent glassy thermoplastics [7,8].

A typical arrangement for interference investigations is schematically depicted in Fig. 7. The fracture mechanics specimen is illuminated in reflection with monochromatic light under normal incidence. A typical fringe system is shown in Fig. 8a for a loaded crack in a PMMA-specimen of high molecular weight. The crack propagation direction is from left to right; two fringe systems are apparent, both with decreasing spacing towards the crack tip: those to the left arise from interference between reflections from the two crack faces, whilst those to the right rise from reflections at the two boundaries between crazed and uncrazed material. In order to calculate the crack openings and craze thicknesses the positions of the individual fringes are determined accurately by scanning the micrographs with a microdensitometer (Fig. 8b) [9,10].

The thicknesses 2v(x) of craze and crack opening are given by basic interference theory

at a bright fringe by $$2v(x) = \frac{\lambda}{2\mu} \cdot (n - 1/2) \tag{10a}$$

and $n = 1, 2, 3 \ldots$

at a dark fringe by $$2v(x) = \frac{\lambda}{2\mu} \cdot n \tag{10b}$$

where n is the order of fringe at position x, λ is the wavelength of monochromatic light and μ is the refractive index.

The crack opening can be determined ($\mu = 1$), if the fringe order n is known. For the calculation of the craze thickness knowledge of the refractive index of the craze and its variation with strain is required. Refractive indices of unloaded crazes have been determined in different glassy polymers by Kambour [11-13] by measuring the critical angle of total reflection of light at the craze/bulk polymer interface; thus e.g. in PMMA the measured craze refractive index was $\mu_o = 1.32$. Under the assumption that the craze extends with complete lateral constraint and using the Lorentz-Lorenz equation, which relates polarizability to the refractive index, the density of the craze material [12] and the craze refractive index μ[13] have been determined as a function of strain. For the evaluation of interference fringe patterns it is more useful to use the numbers of fringes n_l and n_o of the loaded and unloaded craze respectively instead of strain or extension ratio.

Figure 9 shows an example of the experimentally determined shapes (points) of the crack opening and the craze zone in PMMA of high molecular weight. Taking the different scales of the vertical and horizontal axes into

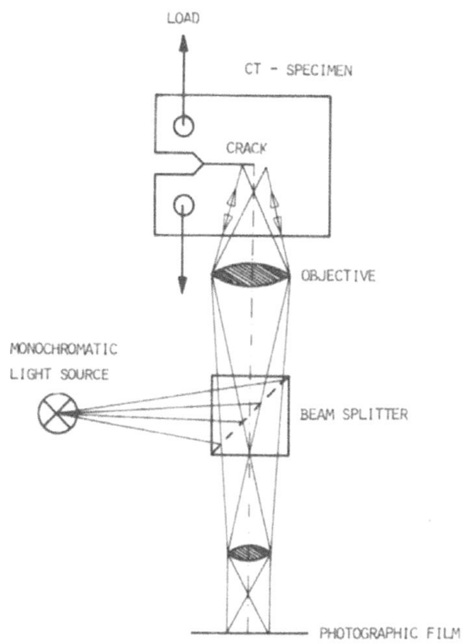

Fig. 7: Sketch of optical interference arrangement for displacement measurements at the crack tip

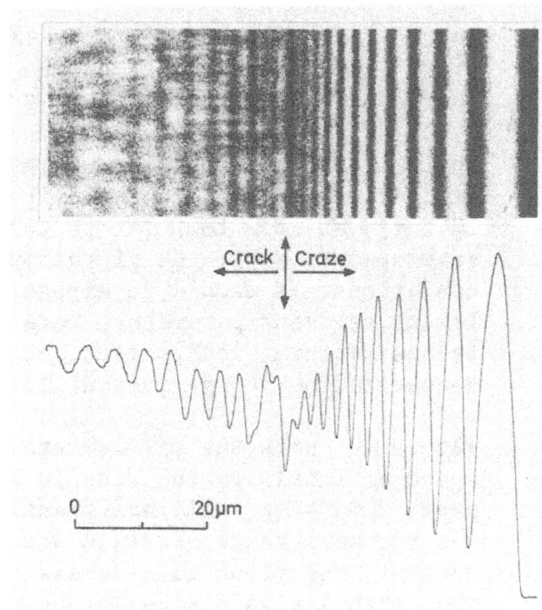

Fig. 8: Interference fringe pattern a) and microdensitometer trace of a crack and craze in PMMA b)

account it becomes evident that the craze is a long thin wedge. In addition the arrangements of molecules and of stretched and broken fibrils are schematically indicated. In Fig. 9 the measured points indicate that the crack tip is blunted and that in this position the craze width 2v is larger than the crack opening.

The similarity between the measured craze zone and the calculated Dugdale plastic zone is apparent when Figs. 6 and 9 are compared.

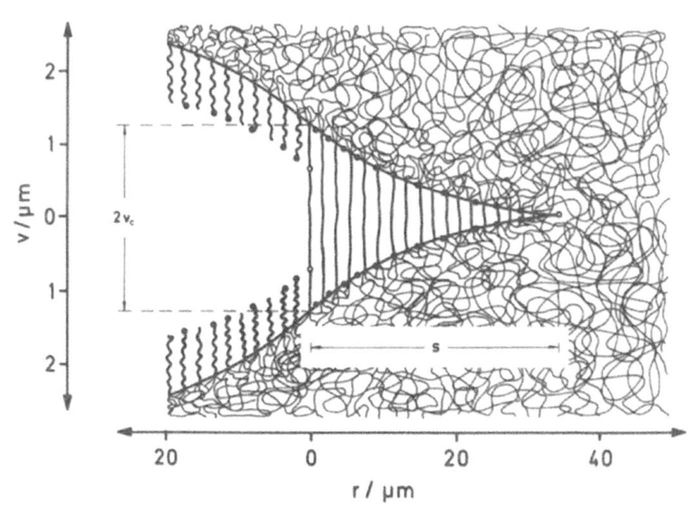

Fig. 9: Measured craze and crack opening (points) in PMMA (Arrangement of molecules and deformation behaviour of the fibrils are schematically indicated)

Some results on craze dimensions at fracture

In the following section it will be demonstrated that the craze zone size at fracture is depending on the loading conditions, temperature and, of course, on the material.

In PMMA of high molecular weight the craze dimensions have been measured at continously moving crack tips in the speed range of 10^{-8} mm/s < $\dot{a}$ < 10^{2} mm/s at room temperature using an experimental set-up especially developed to apply the microscopic interference technique under short time conditions. It should be emphasized that in these experiments the following relevant parameters were measured simultaneously: interference fringe pattern, load, crack length and crack speed. Thus a complete characterization of any instant during the fracture process could be obtained.

Figure 10 shows the maximum craze width $2v_c$ as a function of crack speed $\dot{a}$. Initially the results were interpreted as suggesting that the craze dimensions remained practically constant, independent of crack speed [14]. The mean value obtained was $2v_c = (2.7 \pm 0.2)$ µm. More recent data, however, has prompted a re-examination of these results which reveals that the craze dimensions go through a minimum at speeds of about 10^{-4} - 10^{-3} mm/s. It has been shown that an increase in craze dimensions during crack propagation is to be expected at lower speeds due to a long-time-effect and at higher speeds due to a temperature effect (compare also [15]).

On the other hand in Fig. 11 it can be seen that during fatigue crack propagation (crack growth da per cycle number dN) in PMMA the maximum craze width $2v_c$ is not constant but increases by a factor of nearly 5 (from 0.6 µm to about 2.8 µm) in the predominantly investigated crack propagation range between 10^{-5} and 10^{-2} mm/cycle [16]. It should be mentioned that the plotted curve was obtained at four frequencies in a range from 0.4 to 50 Hz. Figure 11 suggests a lower limit of $2v_c$ at crack propagation rates smaller than 10^{-5} mm/cycle. As has been shown [16,17] an upper limit can be derived by comparing the curve with the results for continuously moving cracks under constant load (Fig. 10) using the crack speed $\dot{a}$ as a common measure.

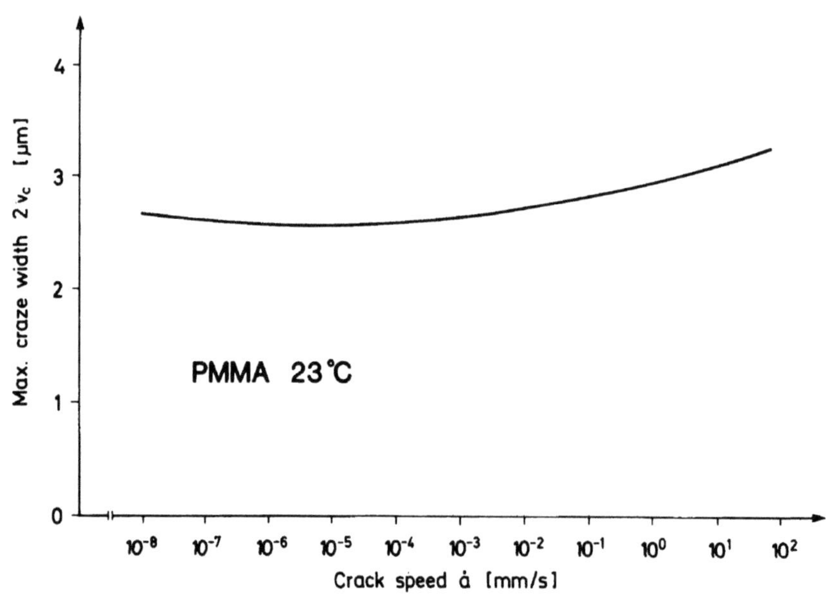

Fig. 10: Maximum craze width $2v_c$ at propagating crack tips in PMMA as function of crack speed $\dot{a}$

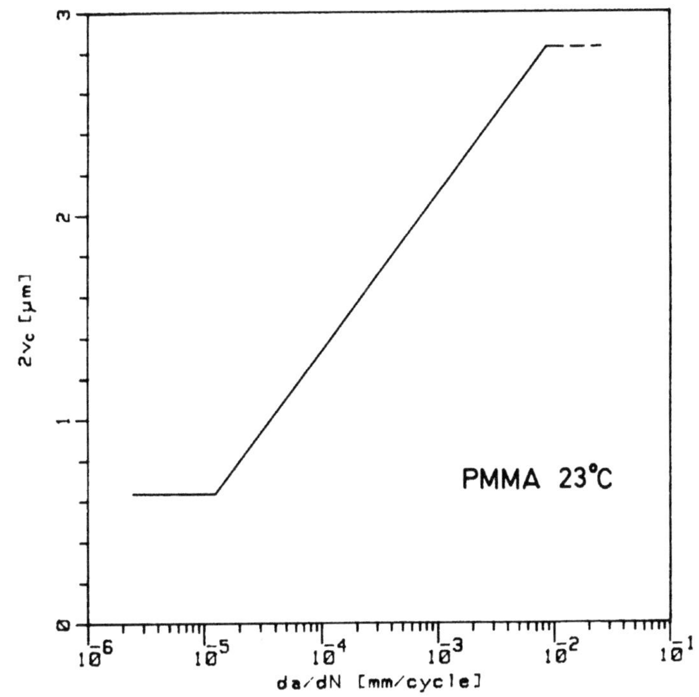

Fig. 11: Variation of maximum craze width $2v_c$ with crack propagation rate da/dN during fatigue crack propagation in PMMA

In Fig. 12 the effect of temperature on the craze size at the crack tip can be estimated from the different fringe pattern in PMMA of high molecular weight. The size of the craze increases with temperature as can direcly be seen from Fig. 12.

In the context of the results for moving cracks discussed above it should be noted that the craze sizes reported now have been measured from just after onset of slow crack propagation to speeds up to 10^{-3} mm/s. Quantitative results of maximum craze width $2v_c$ as a function of temperature T are compiled in Fig. 13 for PMMA [18-20], PC [21] and PVC [18]. At a first glance there seems to be no consistent pattern of behavior. In the two grades of polycarbonate PC (1) (Makrolon, M_w = 20600) and PC (2) (Lexan, M_w = 17000) the maximum craze width $2v_c$ remains constant for each ma-

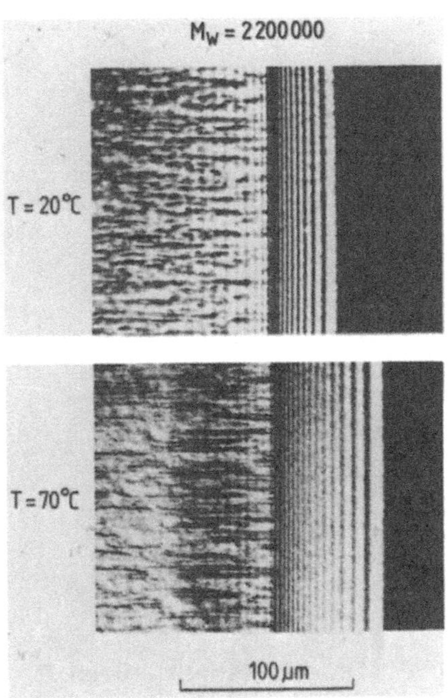

Fig. 12: Interference fringe patterns from the crack tip region in PMMA of high molecular weight at different temperatures

terial over the temperature ranges used; the higher molecualr weight material, PC (1), has the bigger craze width. In three of the four PMMA-grades the maximum craze widths are nearly constant in the measured temperature range and thus show a similar behaviour to PC. Of these materials PMMA (2) is known [18] as a low molecular weight material (M_w = 12000). PMMA (1), which had a high molecular weight (M_w = 2200000), shows a distinct increase in craze width. The molecular weights of PMMA (3) and (4) are not specified. However, from the data in Fig. 13 it may be inferred that these materials were also of low molecular weight.

Two parameters seem to govern the temperature behaviour of the maximum craze width: the relationship between the length of stretched molecular chain and the maximum craze width [10] and the mechanical behaviour of the molecular chain. In the five polymer grades which exhibit a constant $2v_c$ with temperature the length of stretched molecular chain is much smaller than the maximum craze width (= length of stretched fibrils). In the case of high molecular weight PMMA the possible length of a stretched chain is larger than the measured length of stretched fibrils. Thus the high molecular weight PMMA is endowed with a reserve in extensibility of the chains, leading to an increase in stretched fibrils' length with temperature. The temperature behaviour of the stretched fibrils in PVC can be related to the structure of the molecular chain. The chloride side group in PVC is much smaller than that of the ester group in PMMA or the aromatic rings in PC and hence as the temperature increases, the entanglements are more easily loosened under stress leading to the observed decrease in maximum craze width.

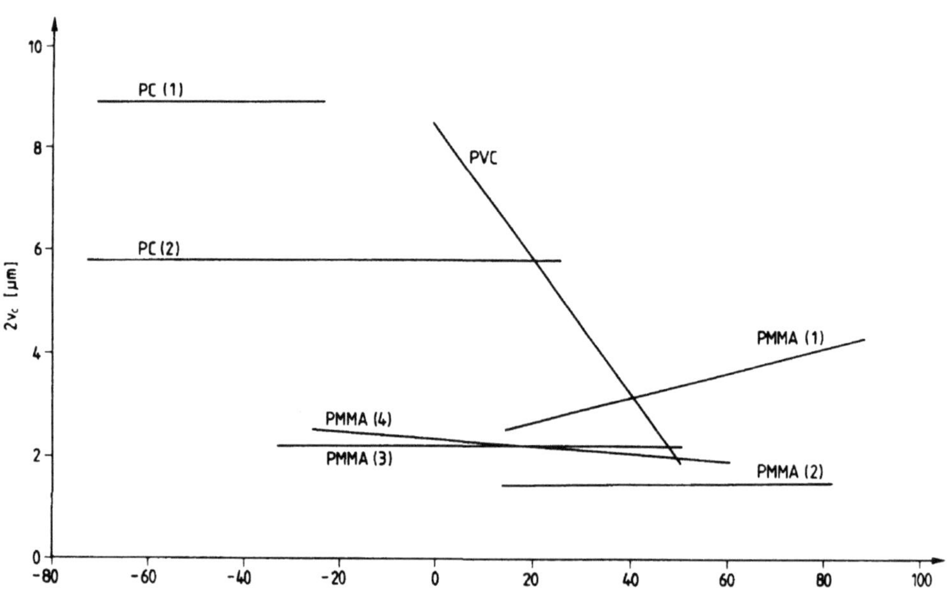

Fig. 13: Maximum craze width $2v_c$ measured at different temperatures T in various polymers

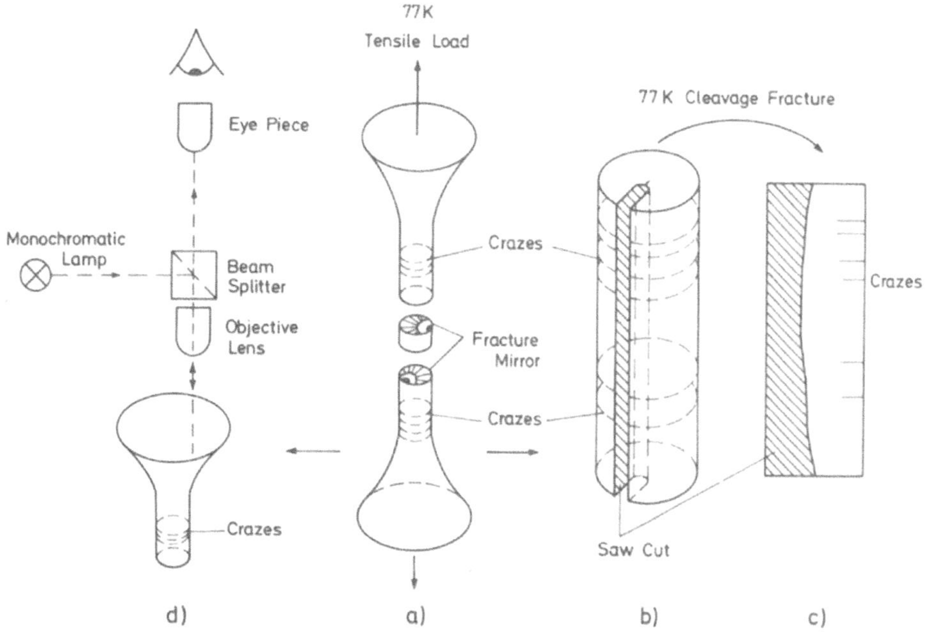

Fig. 14: Sketch of craze investigations in tensile specimens of PC at low temperatures

Crazing of PC at low temperatures

Knowing the importance of crazing in the deformation and fracture behaviour of thermoplastic materials under "normal" conditions the question arises: Does crazing still occur at low temperatures and does it govern the extent of the specimens deformation?

To investigate these problems unnotched specimens of PC have been loaded by uniaxial tensile stresses up to fracture under different strain rates at liquid nitrogen temperature (performed by Hartwig and coworkers [22]). The fragments of the specimens exhibit clearly visible structures in the region near to the specimens surface (see Fig. 14a) and the basic question to be answered is: Are these cracks due to brittle fracture under small deformations or crazes giving rise to a larger potential in deformation of the material?

To decide between the two possibilities interference optics have been used as indicated in Fig. 14d. The fringe pattern recorded is similar to that shown in Fig. 15.[23] however, due the large density of these structures they are harder to discern. Thus, only few photographs were evaluable as shown by the microdensitometer trace in Fig. 16. The evaluation in Fig. 17 gives a shape differing in essential parts from the contour predicted by

Fig. 15: Interference fringe pattern of crazes in PMMA growing from the surface into the interior of the specimen [23]

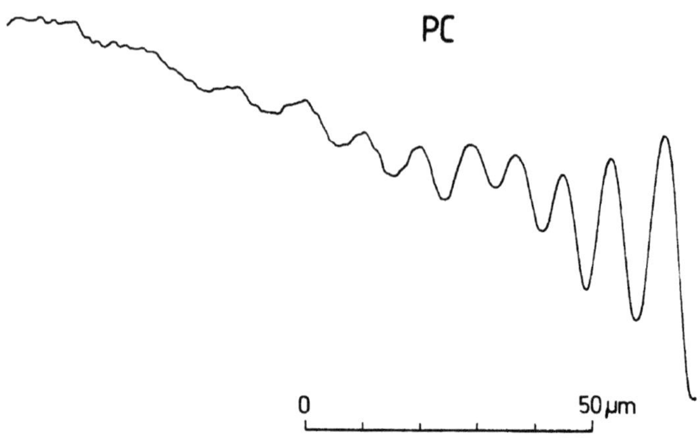

Fig. 16: Microdensitometer trace of the interference fringe pattern from a craze in PC generated at at 77 K

the Dugdale model which describes craze contours. A decision on the nature of this structure from this plot directly, however, is not possible, since the model is only fully valid in the case of a loaded craze when the fibrils are stretched. Taking into account the difference between crack and craze opening a first hint is to be expected. In Fig. 18 the squares of the displacements are plotted. Since in the case of a crack opening a linear course (comp. Eqn. (3)) is to be found it can be concluded that the structure under discussion should be a craze.

To prove this conclusion also SEM-technique has been applied. For this the specimens were handled in a manner depicted schematically in Fig. 15a-c: after groving with a saw and sharpening with a razor blade the specimen was splitted by a wedge perpendicular to the craze system at 77 K. The fracture surfaces were then treated by special preparation method based on low temperature coating with Tantalum atoms. The SEM-photograph (Fig. 19) shows among the fracture markings induced by the splitting fracture parallel lines up to 200 µm length and about 1 µm width which are oriented

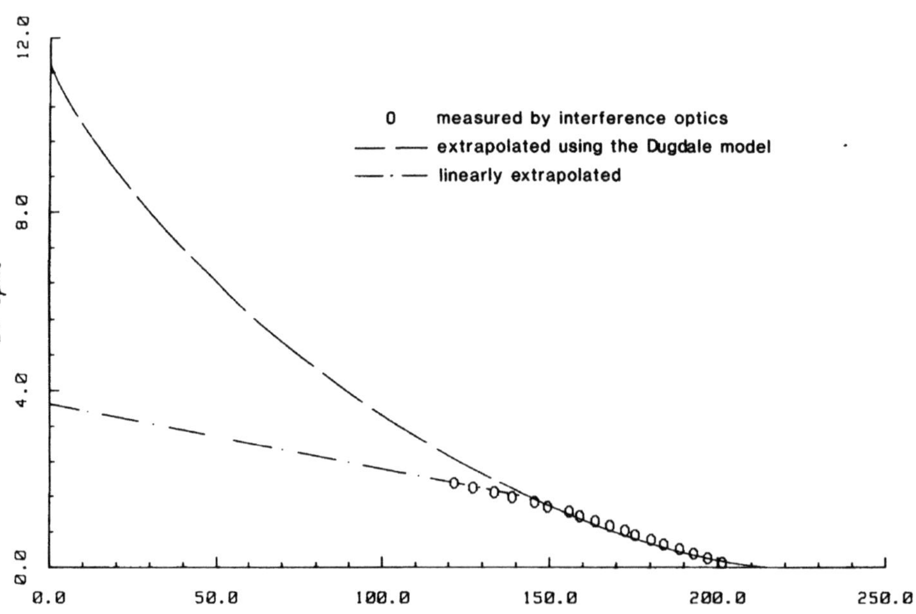

Fig. 17: Shape of a craze zone in PC evaluated from Fig. 16

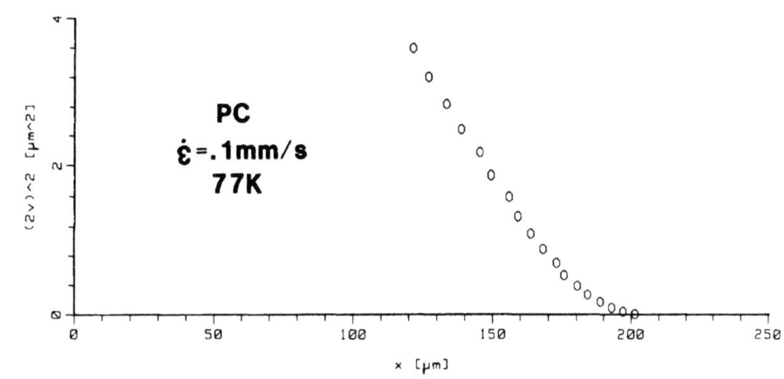

Fig. 18: Square of craze displacement (taken from Fig. 17) as function of the distance from the surface of the specimen

perpendicular to be specimens surface and hence are also perpendicular to the principal stress applied in the original tensile experiment. In Fig. 20 a detail of Fig. 19 is shown under a different visual angle and specimen orientation.

The linear structures in discussion are in Fig. 20 pointed out by arrows. The SEM image contrast clearly shows that the fracture surface along those linear structures is elevated, indicating material popped out of the interior. Since cracks are due to a local deficit of material crazes are domains with increased volume due to their fibrilar structure, thus giving rise to another indication that those linear structures are the traces of crazes.

The final question is, whether these crazes have been generated by thermal stresses during quenching or by the tensile stress at low temperature during the tensile test. This has been proved by investigating similar specimens after multiple quenching without external tensile stresses. Since no hint for crazing in these specimens has been found it can be concluded that crazing takes place under tensile stress conditions even at liquid nitrogen temperature.

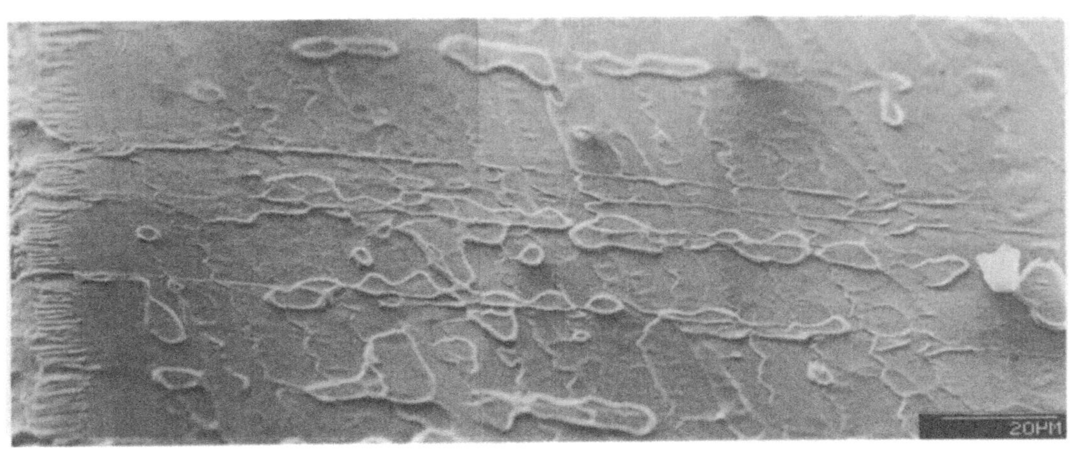

Fig. 19: SEM micrograph of a cleavage fracture surface in PC perpendicular to craze zones produced at 77 K

These investigations are to be seen as a first approach of understandi the deformation and fracture behaviour of thermoplastic materials unde cryogenic conditions. Further results will follow.

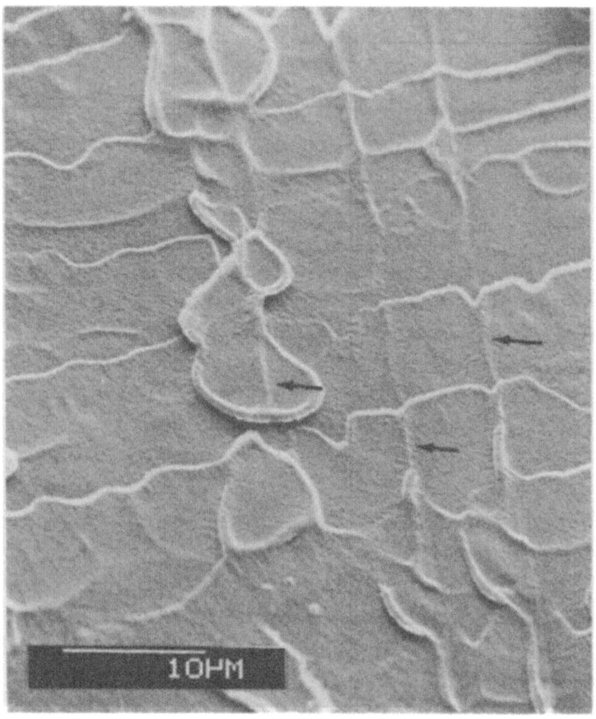

Fig. 20: Marks of craze zones vertical lines) on a cleavage fracture sur face in PC (detai from Fig. 19 turn by 90°)

Acknowledgment

The authors acknowledge partial financial support from the Deutsche Forschungsgemeinschaft (DFG). The assistance of Dr. G. Hartwig and coworkers in performing the tensile tests at low temperature and the encouragment during the investigations is greatly appreciated.

REFERENCES

1. H.H. Kausch, ed., "Crazing in Polymers", Advances in Polymer Science 52/53, Springer-Verlag, Berlin - Heidelberg - New York - Tokyo (1983)

2. M.B. Yaffe and E.J. Kramer, J. Mater. Sci. 16, 2130 (1981)

3. D.S. Dugdale, J. Mech. Phys. Solids 8, 100 (1960)

4. G.I. Barenblatt, Adv. Appl. Mechs. 7, 55 (1962)

5. J.N. Goodier and F.A. Field, in: "Fracture of Solids", D.C. Drucker and J.J. Gilman, eds., John Wiley, New York - London (1963), p. 103

6. J.R. Rice, in: "Fracture - An Advanced Treatise", H. Liebowitz, ed., Academic Press, New York - London (1968), p. 191

7. R.P. Kambour, J. Polym. Sci. A-2 4, 349 (1966)

8. M.I. Bessenov and E.V. Kuvshinskii, Soviet Phys. (Solid State) 3, 1957 (1961)

9. H.R. Brown and I.M. Ward, Polymer 14, 469 (1973)

10. G.W. Weidmann and W. Döll, Colloid and Polymer Sci. 254, 205 (1976)

11. R.P. Kambour, Nature 195, 1299 (1962)

12. R.P. Kambour, Polymer 5, 107 (1964)

13. R.P. Kambour, J. Polym. Sci. A 2, 2165 (1964)

14. M.G. Schinker and W. Döll, in: "Mechanical Properties at High Rates of Strain", Oxford 1979, Conference Series No 47, J. Harding, ed., The Institute of Physics, Bristol-London (1980), p. 224

15. W. Döll, Polymer Eng. Sci. 24, 798 (1984)

16. W. Döll, L. Könczöl and M.G. Schinker, Polymer 24, 1213 (1983)

17. W. Döll, M.G. Schinker and L. Könczöl, in: "Deformation, Yield and Fracture of Polymers", Cambridge 1982, The Plastics and Rubber Institute, London (1982), p. 20-1

18. G.W. Weidmann and W. Döll, Int. J. Fract. 14, R 189 (1978)

19. G.P. Morgan and I.M. Ward, Polymer 18, 87 (1977)

20. R. Schirrer and C. Goett, J. Mater. Sci. Letters 1, 355 (1982)

21. R.A. Fraser and I.M. Ward, Polymer 19, 220 (1978)

22. G. Hartwig and K. Poehlmann, "Low Temperature Fracture Strain at High Deformation Rates", Primärbericht, Institut für Reaktorbauelemente, Kernforschungszentrum Karlsruhe (August 1983)

23. J. Opfermann, "Investigations of Crazing and Fracture of Amorphous Polymers", Institut für Kunststoffverarbeitung der RWTH, Aachen (1978)

A SIMPLE DEVICE FOR SHORT TIME CRYOGENIC TESTS AND ITS APPLICATION IN HIGH VELOCITY TENSILE MEASUREMENTS

K. Pöhlmann and A. Heneka

Nuclear Research Center Karlsruhe, IRB/ZSM

Karlsruhe, Federal Republic of Germany

INTRODUCTION

One undesired property of polymers is their brittleness at low temperatures. Only some linear polymers, e.g. PC, PSU or PE show some ductility even at 4.2 K. On the basis of previous experience it is expected that brittleness increases with increasing deformation rate. Investigations [1] of the strain rate dependence on some selected polymers at 77 K showed that only up to a certain specific strain rate the fracture strain ε_F decreases. At higher strain rates the fracture stress and strain increase remarkably. An explanation of this behavior is given.

The tests were also performed at 4.2 K. Two problems, namely the very time consuming cooling to LHe temperature and the convertion of the available testing machine for cryogenic use, were solved by the development of a simple device, called "mini-cryostat". Its design, operation and the temperature-time behavior during the epxeriment are described. The successful application was demonstrated. Modifications, especially for high velocity tensile tests, are on way.

RATE DEPENDENCE OF FRACTURE STRAIN AND FRACTURE STRESS

In Fig. 1 b the fracture strain ε_F normalized to the static value of the fracture strain ε_{Fo} is plotted versus the strain rate $\dot{\varepsilon}$ for PE,PC and EP. It can be seen that at certain values of $\dot{\varepsilon}$ which are specific of each polymer the fracture strain rises again and exceeds the static values.

This behavior can be explained by stress concentrations due to external loading at the tips of cracks, crazes or voids existing in every polymer specimen and being responsible for the initiation of fracture.

- In the case of very low strain rates some relaxation in the tip zones occurs which reduces the stress concentrations.
- At higher strain rates the deformation time becomes much shorter than the relaxation time for segmential motions of the polymer and there is less stress relaxation. The material becomes more brittle and the ultimate strain decreases.
- At even higher strain rates heat generation by internal friction in the tip zones becomes dominant and leads to an isothermal-adiabatic transition when more thermal power is generated locally than gets diffused.
- At very high strain rates the resulting large temperature rise is ex-

pected to drive the tip zones into temperature ranges where plastic deformation occurs and the fracture strain increases caused by a brittle-tough transition.
The initiation of macroscopic fracture is thought to depend on stress concentrations at the crack tips. These stress concentrations can be relaxed to a relatively large amount by a plastification of the tip zones due to a brittle-tough transition which is controlled by the damping behavior specific of each polymer. This leads to a more homogeneous distribution of the load applied and therefore the ultimate values of strain and stress increase. As shown in Figure 1a, the fracture stress is nearly constant up to the specific strain rates after which a steep rise is observed similar to the rise of the fracture strain.

It is seen from Figure 1 that PE has its minimum in fracture strain at the lowest value of strain rate according to its relatively lowest transition temperature. The slight increase for PC reflects its low but very broadly distributed damping peak. The somewhat higher transition temperature of PC is expressed by a little higher value of the strain rate, $\dot{\varepsilon}$, at the minimum of fracture strain. EP has a sharp damping peak, similar to PE, at a relatively high temperature. Therefore, there is a steep increase of the fracture strain at the relatively highest value of $\dot{\varepsilon}$. Thus the fracture behavior at very low temperatures is controlled by secondary phase transitions taking place at much higher temperatures, the link being an adiabatic temperature rise at the tips of microcracks. Investigations at 4.2 K could confirm the explanation of the strain rate behavior.

MATERIALS

The polymers investigated were selected with respect to the different temperatures of their secondary dispersion regions which cause brittle-tough transitions due to segmential motions in the main chains. Polyethylene (PE) has a high secondary maximum at 160 K. Polycarbonate (PC) shows a low but very broad damping peak at 180 K with a halfwidth of $\pm$ 60 K. The damping behavior was measured at a frequency of about 12 Hz[2].

APPARATUS

High Velocity Tensile Measurements

The tests were performed with a special hydrodynamic tensile machine within the range of cross-head speeds of 10^{-4} to 8.5 m/s. This corresponds to strain rates from 5×10^{-3} to 5×10^{2} s^{-1} due to the geometric dimensions chosen. An impact bar was accelerated and connected with one clamp of the specimen over a push rod after having attained a constant speed.

The fracture strain was determined by the cross-head displacement which was intercalibrated with the strain of the specimen by small strain gauges. This calibration has to be performed for each material, temperature and velocity applied. The calibration factor was examined to be independent of the cross-head speed up to 0.3 m/s. For higher speeds the experimental setup has to be improved electronicly. The change of load as a function of deformation time was detected by a piezo-quartz.

To distinguish whether there is impact or nonimpact loading reference is made to the considerations of R.D. Spangler[3]. When a specimen is loaded at one end while the other is clamped, a stress wave travels along the specimen at the velocity of sound for that material. The wave is reflected at the fixed end and travels along the specimen several times. After 10 passages one may assume that the stress wave has damped out and the specimen is under uniform tension from one end to the other (therma-

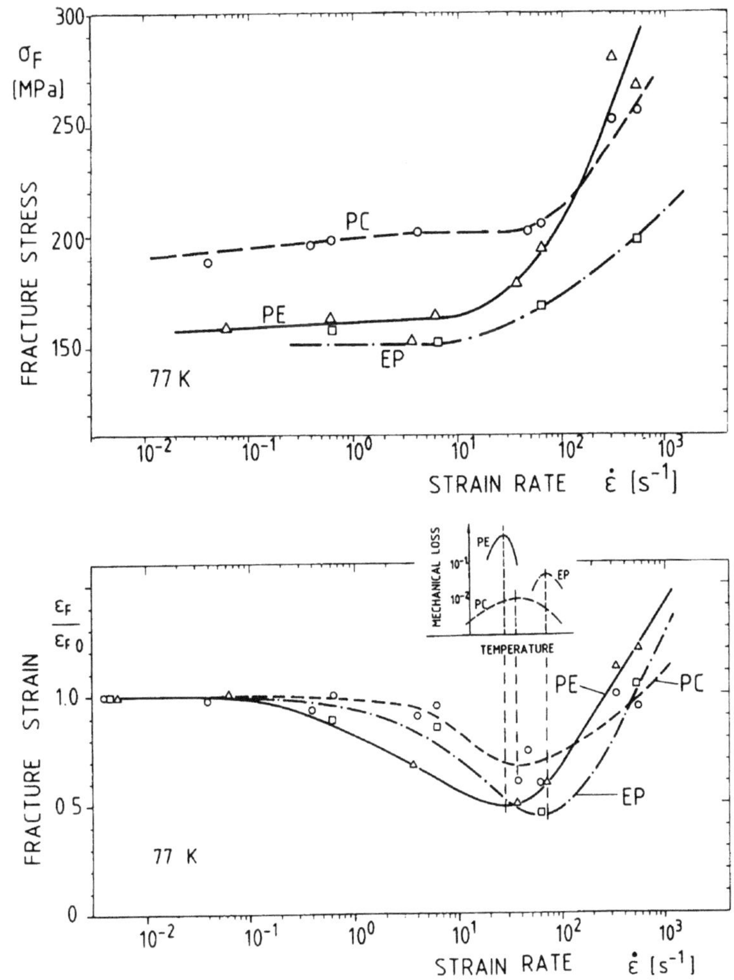

Fig. 1. a) Fracture stress versus strain rate for different polymers

b) Fracture strain, ε_F, normalized to the static ε_{Fo} versus strain rate

lisation). This is the case for nonimpact loading and one better speaks of high velocity tension. Real impact loading would be present only if the time to fracture were less than the time for one passage of the stress wave so that the specimen breaks due to the high stress-induced strains at the impacted end while the strains at the other end are zero. Even for the shortest times to fracture and the highest cross-head speeds at least ten passages for a stress wave were obtained for all materials tested. For this reason the experiments are comparable in the whole strain rate range and can be designated as nonimpacting or high velocity tensile test.

The testing machine available for tests at the high velocities (up to 8.5 m/s) is designed for RT only. There were no problems to transform the machine for experiments at 77 K but extensive arrangements for those at 4.2 K are needed. Moreover, the pure testing times are very short but the time of cooling to LHe temperature is unreasonably long so that those experiments are rather uneconomic. Therefore, a simple device, the "mini-cryostat", has been developed allowing to perform measurements at 4.2 K in short time tests.

Principle of Measuring

The application of the mini-cryostat in tensile tests and the experimmental procedure are shown schematically in Fig. 2. The specimen is mounted at room temperature (RT) into the mini-cryostat and then precooled to 77 K in a small Dewar vessel. Thereafter, it is transposed into a helium cryostat. After having reached the temperature of 4.2 K the minicryostat, filled with liquid helium, is removed and fastened in the testing machine. This is possible within a few seconds by a simple hang-in mechanism. The tensile test is then performed and the mini-cryostat can be reheated and remounted afterwards. The main advantage of the method described is that the testing machine can be left at room temperature and only the small mini-cryostat has to be cooled down to 4.2 K.

Design of the Mini-cryostat

The design of the mini-cryostat is shown in Fig. 3. Essentially it consists of three parts: the upper part composed of a clamp and a bolt for the upper hang-in mechanism; the lower part composed of the second clamp, the bottom and the fork for the lower hang-in mechanism; and a GFRP-tube. This tube is fastened and sealed with an adhesive tape to the bottom so that the mini-cryostat can be filled with a coolant. The open upper side is covered with superinsulation and Teflon foils. Additionally, superinsulation and Teflon foils at the tube and the bottom provide a good insulation of the mini-cryostat.

A teflon cylinder for fixation between the clamps protects the small specimens from being bent during handling. The fork of the upper hang-in mechanism is screwed into the holding bar and establishes a stiff connection with the machine. The bolt at the upper part of the mini-cryostat can be hooked easily, and, likewise, the bolt in the push rod in the lower part of the hang-in mechanism can be hooked into the fork of the mini-cryostat. The simple hang-in mechanism permits the mounting of the mini-cryostat into the machine after removing it from the helium cryostat within roughly 10 to 15 seconds. All parts were formed on a lathe. This guarantees an exact centring in the axial direction.

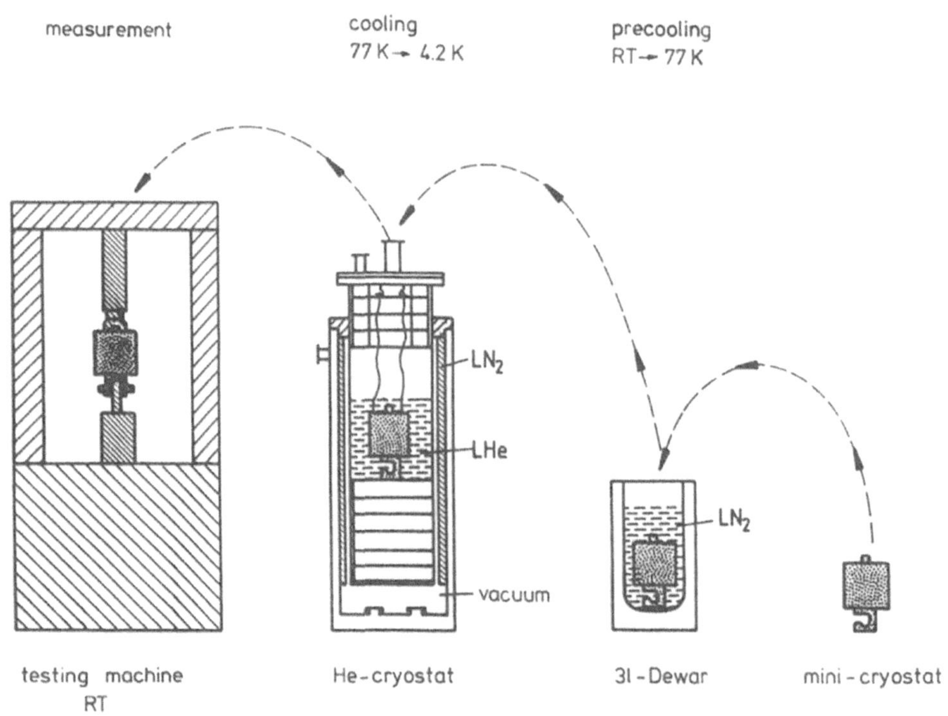

Fig. 2. Principle of measuring

Time-Temperature Behavior

It is of special interest whether the total time necessary for the whole experiment (removing the mini-cryostat from LHe, fastening in the machine and tensile test) is short enough so that the temperature of the specimen does not exceed 4.2 K. Therefore, the temperature on the surface of the specimen was determined with a silicon temperature sensor in several tests. Figure 4 shows a plot of the measured temperatures versus time starting from the point after removing the mini-cryostat out of liquid helium. The decisive result is that all tests yielded a lower limit of 20 seconds. Within this range of time the temperature remains constant at 4.2 K and the measurement will be finished with certainty (the test times are of the order of milli-seconds up to one or two seconds). Accordingly the assumption that the temperature is kept on 4.2 K during the tensile test, is correct. The pre-examinations demonstrated the applicability of this method for measuring at cryogenic temperatures.

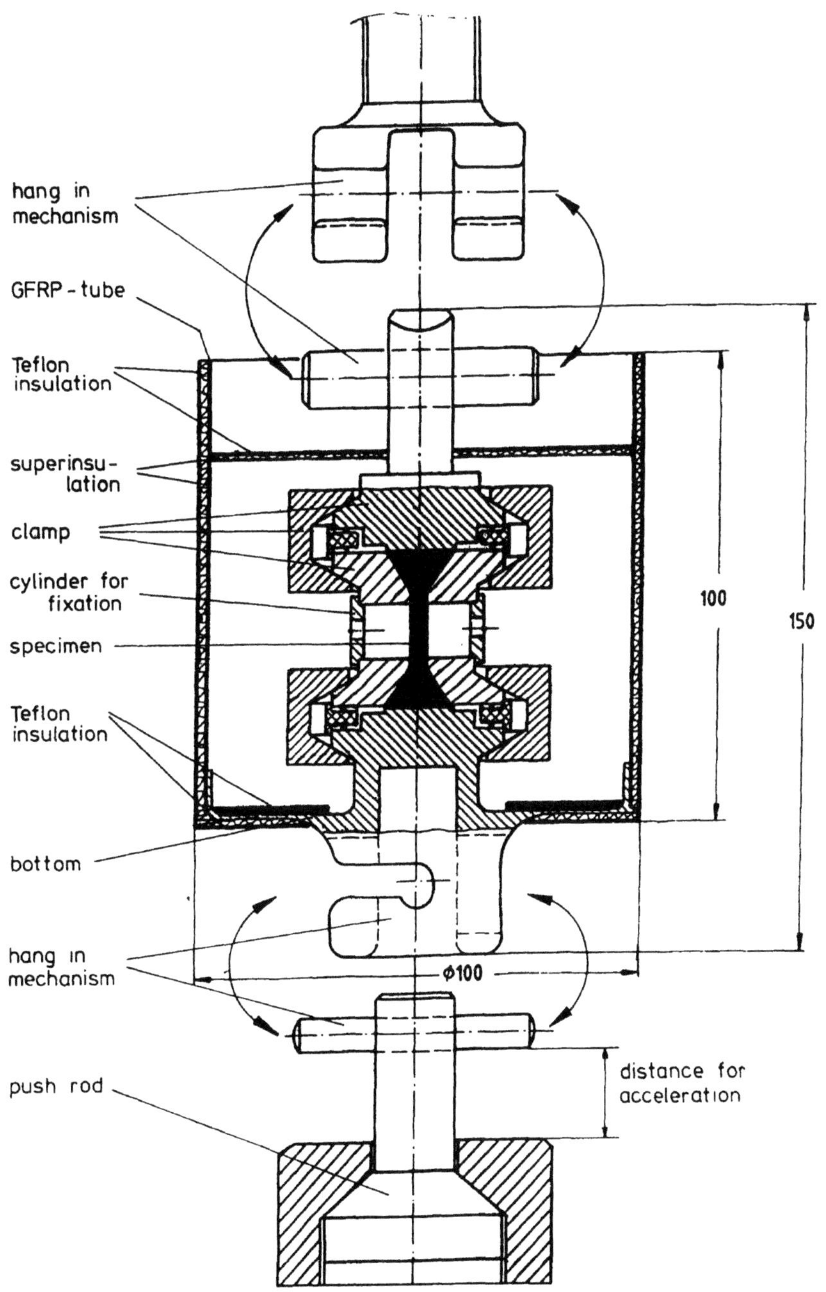

Fig. 3. Design of the mini-cryostat

The periods of precooling from RT to 77 K and cooling down from 77 K to 4.2 K were also determined in several tests and found to be about 4 to 4 1/2 minutes and 3 to 4 minutes, respectively. It is expected that in the experiment 5 minutes are always on the safe side. Reheating and repeated mounting of the mini-cryostat last about 10 minutes. All these periods give a total measuring cycle of about 20 minutes duration represented schematically in Fig. 5. This relatively short time for one cycle allows a great number of measurements to be made at 4.2 K. In one day, for example, 12 - 15 tensile tests can be performed because of the easy handling and the low expenditure in terms of cryoengineering.

Economy

It was experienced that only 6 to 8 l LHe are consumed in one measurement which is due to the good insulation, the small geometric dimensions and the good precooling to 77 K. This small amount of helium is not recoverable as in experiments with large cryostats. However, the costs of these losses are exceeded by the liquifying costs and the costs of the losses in large cryostats where about 100 to 120 l LHe are needed for one measurement. So the method presented is rather economic since it saves not only experimental time but also cooling costs.

RESULTS OBTAINED AT 4.2 K

Preliminary measurements were performed at 4.2 K using the mini-cryostat described. The method itself has proved true for cryogenic experiments. Some results were obtained up to a cross-head speed of 0.1 $\frac{m}{s}$ only. At cross-head speeds equal to or lower than 0.1 $\frac{m}{s}$ pure linear load-time diagrams were recorded. However, at somewhat higher speeds the curves were falsified by the superposition of vibrations with a specific frequency. At even higher speeds the amplitude of these vibrations becomes so high that the diagrams can not be analysed any longer.

There are mainly two ways to eliminate the influence of the vibrations to the load-time diagram. The first is to place the load cell closer to

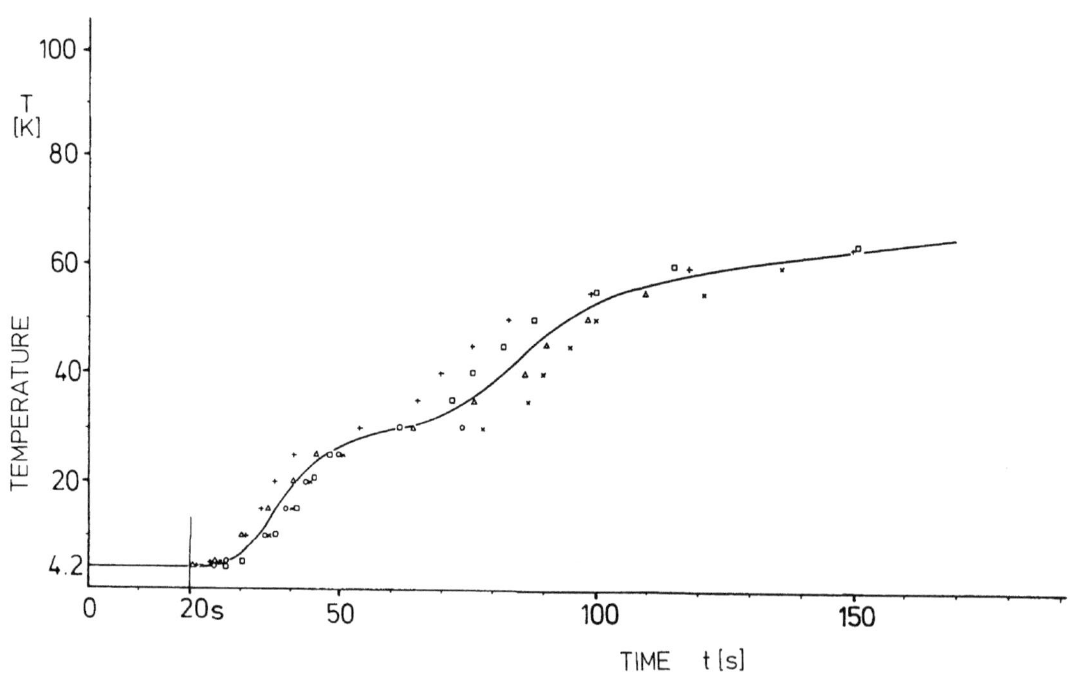

Fig. 4. Time-temperature behavior of the mini-cryostat

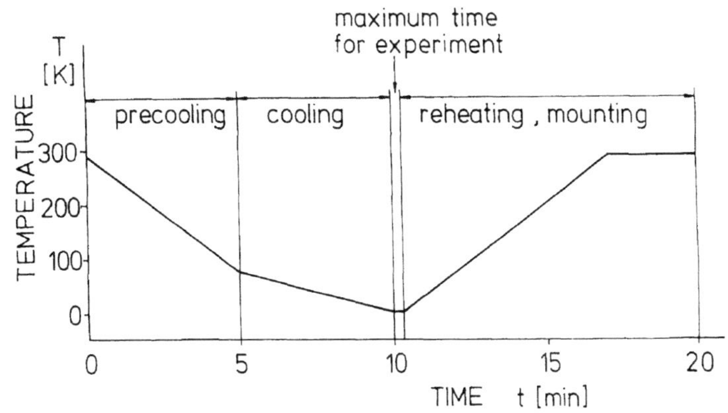

Fig. 5. Measuring cycle (schematically)

the specimen which could be of success if the vibrations stem from parts of the machine. Minimizing the massof the mini-cryostat is the second possibility if the mini-cryostat or the mini-cryostat/push rod system is the origin of the vibrations. Modifications of the mini-cryostat are on way to obtain reasonable results at high velocities.

ACKNOWLEDGMENT

The authors would like to thank Dr. G. Hartwig for valuable discussions and advices. Parts of the measurements were performed at a test facility of the "Fraunhofer Gesellschaft" Berghausen, which is greatly acknowledged.

REFERENCES

1. G. Hartwig and K. Pöhlmann, Low temperature fracture strain at high deformation rates, in "Advances in Cryogenic Engineering", Vol. 30, A.F. Clark and R.P. Reed, eds., Plenum Press, New York (1984), pp. 83 - 88

2. G. Hartwig and G. Schwarz, Correlation of dielectric and mechanical damping at low temperatures, ibidem, pp. 61 - 70

3. R. D. Spangler, Basic design features of a nonimpacting, pneumatically driven, hydraulically damped high speed tester, J. Appl. Polym. Sci., Vol. VI, No. 21, 264 - 268 (1962)

APPARATUS PERMITTING RAPID CONSECUTIVE

FLEXURAL TESTING AT 4.2 K

D.Evans, R.Luckock and J.T.Morgan

Rutherford Appleton Laboratory
Chilton, Didcot, England

INTRODUCTION

Testing materials at low temperatures is a time consuming and expensive procedure. For this reason, it is understandable that many research programs are limited in scope, both in the number of replicates, and in the number of materials that are examined in detail.

The main difficulties are the problems associated with specimen changing (particularly if strain measurement on the specimen is needed), and the necessity to warm the straining assembly to near ambient to change the specimen.

The apparatus described in this report was designed for testing plastics and composites in flexure. It attaches to a compression assembly mounted below the crosshead of an Instron testing machine and is immersed in liquid helium contained in a cryostat mounted on the machine.

In flexural testing, relatively small loads produce large specimen deflections so that strain measurement on the specimen is not necessary, the load/extension data from the testing machine being used to derive stress/strain curves.

The flexural test specimen is inserted through the top of the cryostat and is manoevered into the testing position by a crosshead movement. After test the specimen is rejected, again by crosshead motion. The number of tests that may be consecutively performed is thus limited only by the available space at the bottom of the cryostat for the used specimens.

DESCRIPTION

The testing fixture is shown diagrammatically in Figure 1 which attempts to show the three main positions:-

1) Specimen loading
2) Testing position
3) Specimen rejection after test

With the central tension member at its lowest point, the swinging

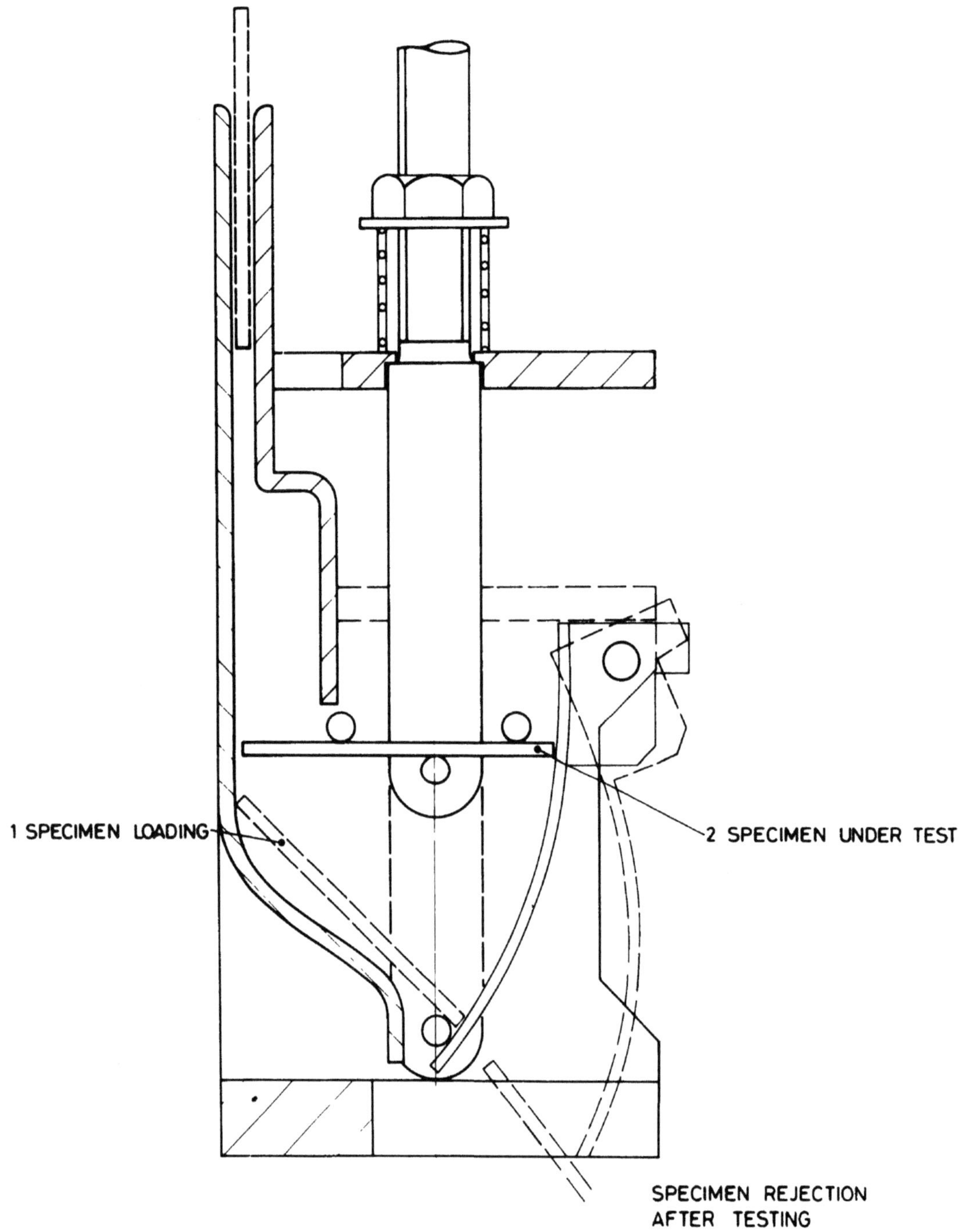

Fig. 1 Diagram of testing fixture showing the three main positions.

Fig. 2. Apparatus with central tension member at its lowest point. Guide arm in position to receive a specimen.

Fig. 3. Apparatus with tension member at highest point.

specimen guide arm is automatically held in position to receive the specimen which is dropped vertically down a chute. The guide prevents the specimen passing completely through the testing assembly. Lowering the crosshead raises the central tension member which automatically positions, with the help of the guide, the specimen into the testing position.

After allowing sufficient time for the specimen to attain the test temperature, the test is completed. With further crosshead motion, the guide is swung out of position by the action between a peg on the guide arm and a cam on the tension member. When the tension member is lowered, the specimen falls to the bottom of the cryostat. The tension member is then further moved to its lowest position which moves the guide arm in position to receive the next specimen. Specimens that are bent but not broken during test are also rejected by this procedure.

Figure 2 shows the apparatus with the central tension member at its lowest point with the guide arm in position to receive a specimen. Figure 3 shows the tension member at its highest point with the guide arm being swung so that the specimen will be ejected when the tension member is lowered.

The simplicity of operation and the minimum of moving parts means that the apparatus does not become immobilised by small amounts of air admitted to the cryostat with the specimens.

SUMMARY

This apparatus has permitted the rapid consecutive testing of plastics and composites in flexure at cryogenic temperatures. It has resulted in significant savings of both testing time and liquid helium consumption.

RADIATION EFFECTS ON POLYBUTYLENETEREPHTHALATE RESINS AT LOW TEMPERATURE

Hitoshi Yamaoka, Kiyomi Miyata, Yoshio Nakayama, and Hiroyuki Yoshida

Research Reactor Institute
Kyoto University
Kumatori, Osaka, Japan

ABSTRACT

The mechanical properties of irradiated polybutyleneterephthalate (PBT) resins were measured by means of flexural tests. The specimens irradiated at ambient temperature showed no appreciable change both in flexural strength and in flexural modulus. When the specimens were irradiated at 20 K and warmed up to 77 K after irradiation, only a slight mechanical change occurred in mica flake reinforced PBT, whereas remarkable reduction of flexural strength was observed in glass fiber reinforced PBT. Observation on fracture surface in irradiated glass fiber reinforced PBT with scanning electron microscope revealed a marked debonding in the matrix-filler interface.

INTRODUCTION

Problems on the stability and the durability of organic insulators in intense radiation field at low temperatures are of great importance for designing superconducting magnets of fusion reactors.[1]

Extensive studies on radiation effects of organic insulators at cryogenic temperatures have been carried out in recent years.[2-10] However, almost all experiments were made only with a few materials such as epoxy and polyimide resins. Polybutyleneterephthalate (PBT) resins are widely used as a typical engineering plastics owing to the good electrical properties of matrix polymer and to the ease of industrial processing.

In the present study, the changes in mechanical propertis of PBT resins irradiated at low temperature have been measured by means of flexural tests. The scanning electron microscopic observation has also been performed on fracture surfaces of irradiated PBT resins.

EXPERIMENTAL

Materials

The polymer matrix was commercially available PBT (Novadur 5010, Mitsubishi Kasei Kogyo Co., Ltd.). Mica flake (Suzorite mica 325S,

Marietta Resources Intern. Corp.) and glass fiber (CS6E-231, Nitto Boseki Co., Ltd.) were used as fillers of composites. The average sizes of mica flakes were 40 μm in diameter and 1 μm in thickness. The length and the diameter of glass fibers were approximately 450 μm and 9 μm, respectively. The mica flake reinforced PBT (M-PBT) and the glass fiber reinforced PBT (G-PBT) were prepared by the injection molding method at 523 K and subsequently postheated for 3 h at 373 K. Conventional silane coupling agents were used to improve the interfacial adhesion strength between the fillers and the matrix. The content of the fillers was 30% by weight. Test specimens provided from the molded plates were cut parallel or transverse to the flow direction in the injection molding. Dimensions of the specimens in mm were 2 x 10 x 40.

Irradiation and Tests

Irradiations of the specimens were carried out by using the low temperature irradiation facility of Kyoto University Reactor. The average temperature of the facility is 20 K under reactor operation of 5 MW. The neutron energy spectrum in MeV region is very near to that of fission spectrum on ^{235}U ($f_{max} \simeq$ 2 MeV). The maximum values of fast neutron flux and γ-ray dose rate at the top position of the facility are $2.5 \times 10^{11} n/cm^2/s$ and 1.2×10^7 R/h, respectively. After irradiation, the specimens were kept at 77 K immersed in liquid nitrogen in order to wait the decay of total radioactivity to a safe handling level. Irradiations at ambient temperature (about 350 K) were performed in the same facility without cooling.

The mechanical properties of the specimens were measured by means of flexural test with a Shimadzu Autograph Testing Machine. Three-point bending configuration with a span length of 25 mm was employed at a crosshead speed of 1 mm/min. The tests for all specimens irradiated at 20 K were performed at 77 K without warming up to room temperature. The flexural modulus of the specimens was calculated from the initial slope of load-deflection curve and the flexural ultimate strength was also determined. Data of flexural

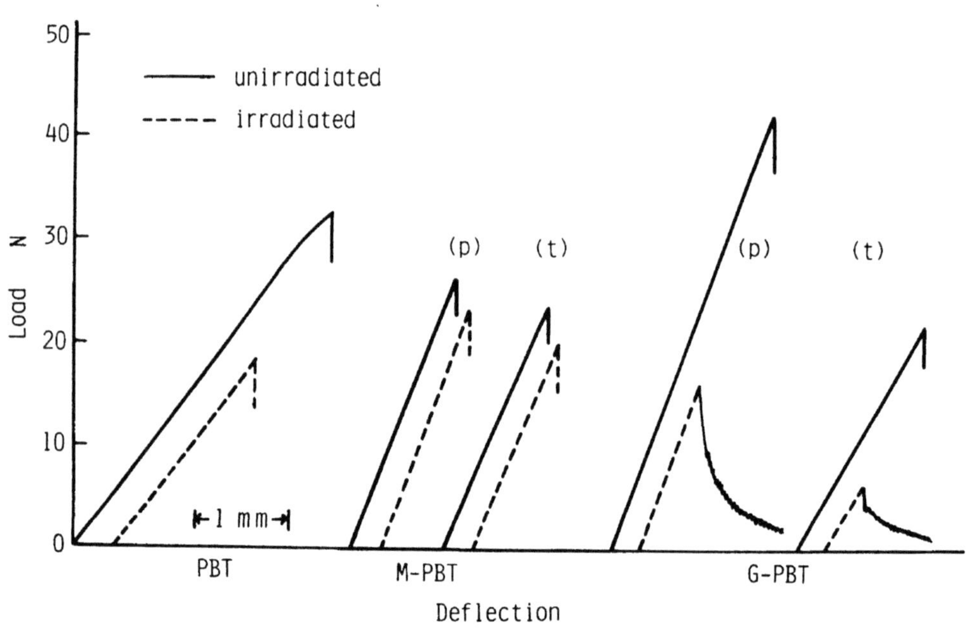

Fig. 1. Typical load-deflection curves of unirradiated and irradiated PBT, M-PBT, and G-PBT at 77 K.

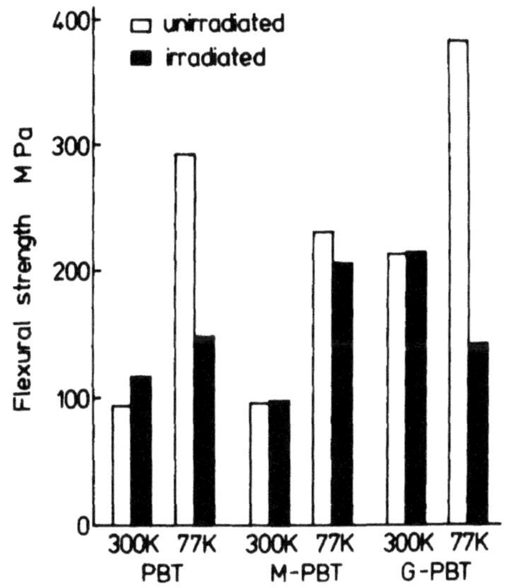

Fig. 2. Effect of reactor irradiation on flexural strengths of PBT, M-PBT, and G-PBT.

strength and modulus were obtained from five specimens for each irradiation condition.

After complete bending, the fracture surfaces of irradiated specimens were observed with a JXA-50A scanning electron microscope (Japan Electron Optics Laboratory Co., Ltd.) operated at 50 kV.

RESULTS AND DISCUSSION

Mechanical Tests

No essential change in the flexural strengths of all specimens was observed for irradiation up to fast neutron fluence of 2.1×10^{11} n/cm^2 and γ-ray dose of 5.7×10^7 R at 20 K. Over this irradiation limit, the flexural strengths of the specimens almost linearly decreased with increasing the irradiation dosage. Typical load-deflection curves of irradiated PBT, M-PBT, and G-PBT are shown in Fig. 1, along with those of unirradiated specimens. Here, the symbols of (p) and (t) represent the specimens cutting parallel and transverse to the flow direction in the injection molding. Most striking features in Fig. 1 are the remarkable reduction of flexural strength and the serration mode after breaking point observed in irradiated G-PBT. These will be discussed later.

In the present experiment, the specimens were exposed to reactor irradiation at 20 K up to highest fast neutron fluence of 8.6 $\times 10^{15}$ n/cm^2 and γ-ray dose of 4.0×10^8 R. The changes in flexural strengths of the (p) specimens irradiated at the highest dosage are shown in Fig. 2. For comparison, the results of the same dosage irradiation at ambient temperature are also added in Fig. 2, where the specimens were irradiated at about 350 K and measured at 300 K. Although the strength of M-PBT is lower than that of G-PBT in the unirradiated condition, the former shows a much higher tolerance for irradiation. In the case of G-PBT, over 60% reduction of the strength, which is lower than that of the matrix itself, was observed. This apparently indicates that the debonding between the fiber and the matrix occurs in the irradiated composites. On the other hand, no appreciable change was found in all irradiated specimens at ambient temperature.

Fig. 3 shows the Flexural moduli of PBT, M-PBT, and G-PBT ir-

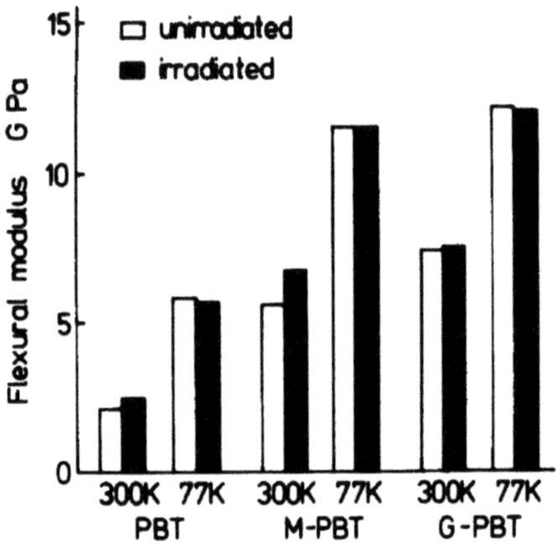

Fig. 3. Effect ofreactor irradiation on flexural moduli of PBT, M-PBT, and G-PBT.

radiated at cryogenic and ambient temperatures. In contrast to the results of flexural strength, only a little change was seen in all specimens both at cryogenic and at ambient irradiations. Similar results have been obtained in irradiated G-10CR and G-11CR by Coltman, Jr. and Klabunde.[6] They reported that the flexural modulus of the specimens was substantially less sensitive to radiation than the flexural strength. These results indicate that the effect of radiation does not weaken the bond strength between fillers and matrices, but reduces the area of interaction surface between fillers and matrices. The reason of greater radiation tolerance of M-PBT would be attributed to a large percentage of reinforcement for a given sectional area due to flatness of mica flake.

The mechanical properties of unirradiated and irradiated specimens in the two orthogonal directions at 77 K are summarized in Table 1. As shown in our previous report for unirradiated PBT resins,[11] G-PBT shows strong anisotropic properties in different force directions since glass fibers in composites are usually aligned parallel to the flow direction in the injection molding. A similar situation is observed in irradiated G-PBT. The notable feature in the (t) specimens of G-PBT is the occurrence of higher radiation damage in the mechanical properties, compared with the irradiated (p) specimens. The values obtained for M-PBT exhibit less anisotopy than G-PBT, reflecting the two dimensional reinforcement in the plane of mica flake.

Scanning Electron Microscopic Observation

Photographs given in Fig. 4 represent the fracture surfaces of irradiated PBT resins observed with scanning electron microscope. In the case of M-PBT specimens shown in Fig. 4 (a) and (b), little difference between ambient and cryogenic irradiations was seen in the appearance of the fracture surfaces and the failure mode. This is consistent with the greater radiation tolerance found in the mechanical tests. The good adhesive nature with the matrix and the efficient reinforcement by overlapping of mica flake are still maintained even after cryogenic irradiation. A small decrease in the flexural strength of M-PBT irradiated at 20 K is attributed to the radiation-induced embrittlement of the matrix itself.

On the other hand, the photograph of G-PBT irradiated at 20 K (Fig. 4 (d)) is much different from that of G-PBT at 350 K (Fig. 4 (c)). The surface of glass fiber in G-PBT irradiated at 20 K is

Table 1. Mechanical properties of unirradiated and irradiated specimens in the two orthogonal directions at 77 K

Sample	Flexural strength (MPa)				Flexural modulus (GPa)			
	Unirradiated		Irradiated		Unirradiated		Irradiated	
	Average	σ*	Average	σ*	Average	σ*	Average	σ*
PBT	292	13.7	149	26.8	5.8	0.04	5.7	0.13
M-PBT (p)	232	6.5	207	7.3	11.5	0.29	11.5	0.23
M-PBT (t)	213	5.0	176	3.6	10.9	0.14	10.5	0.31
G-PBT (p)	383	3.5	144	12.4	12.2	0.22	12.1	0.33
G-PBT (t)	192	7.7	52	4.6	8.0	0.12	7.0	0.78

* : Standard deviation

(a) M-PBT irradiated at 350 K.

(b) M-PBT irradiated at 20 K.

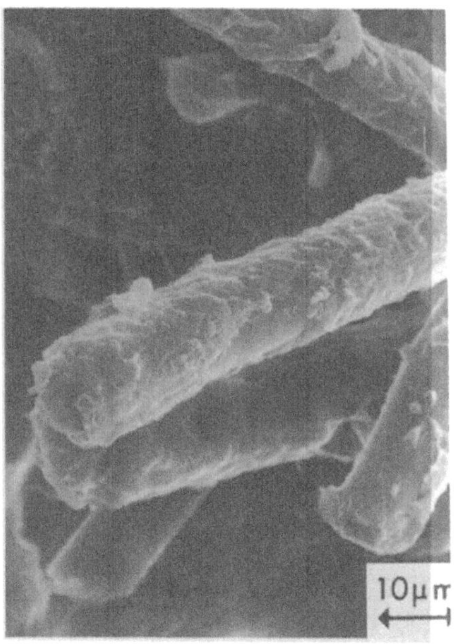

(c) G-PBT irradiated at 350 K.

(d) G-PBT irradiated at 20 K.

Fig. 4. Scanning electron micrographs of fracture surfaces in irradiated M-PBT and G-PBT.

very smooth, indicating the debonding with the matrix. The drastic decrease in flexural strength of G-PBT at 20 K arises from this obvious debonding. The serration mode of irradiated G-PBT shown in Fig. 1 can be corresponding to the phenomenon that the debonded fibers are pulled gradually from the matrix phase after the breaking. Similar observations on debonding have recently been reported in relation to the radiation damage of epoxy resins.[8,10] These results indicate that the surface treatment of fillers and the choice of coupling agents are essential to the composites used in intense radiation field at low temperature.

CONCLUSIONS

In the flexural tests of PBT resins irradiated at 20 K, the strength of M-PBT was less sensitive to radiation than that of G-PBT, whereas the moduli of both specimens were almost unchanged during the course of irradiation. These results indicate that the effect of radiation decreases the area of interaction surface between fillers and the matrix. The greater radiation resistant property of M-PBT is attributed to the efficient overlapping of mica flake for reinforcement. The drastic reduction in strength of G-PBT arises from a marked debonding in the matrix-filler interface. The observations of fracture surfaces of irradiated PBT resins gave additional evidence for the difference in radiation damage between M-PBT and G-PBT.

ACKNOWLEDGEMENTS

The authors are grateful to Kuraray Co., Ltd. for a generous gift of the test specimens. This work is partly supported by Grant in Aid for Scientific Research No. 57055030, Ministry of Education in Japan.

REFERENCES

1. F. W. Clinard, Jr. and G. F. Hurley, Ceramic and organic insulators for fusion applications, J. Nucl. Mater. 103/104: 705-716 (1981).
2. R. R. Coltman, Jr., C. E. Klabunde, R. H. Kernohan, and C. J. Long, "Radiation Effects on Organic Insulators for Superconducting Magnets," ORNL/TM-7077, Oak Ridge National Laboratory, Oak Ridge, Tennessee (1979).
3. S. Takamura and T. Kato, Effect of low temperature irradiation on the mechanical strength of organic insulators for superconducting magnets, Cryogenics 20: 441-444 (1980).
4. R. R. Coltman, Jr. and C. E. Klabunde, Mechanical strength of low-temperature-irradiated polyimides: A five-to-tenfold improvement in dose-resistance over epoxies, J. Nucl. Mater. 103/104: 717-722 (1981).
5. S. Nishijima, S. Ueta, and T. Okada, The effects of low temperature irradiation on the cryogenic fatigue resistance of epoxy resin used in superconducting magnets, Cryogenics 21: 312-313 (1981).
6. R. R. Coltman, Jr. and C. E. Klabunde, The strength of G-10CR and G-11CR epoxies after irradiation at 5 K by gamma rays, J. Nucl. Mater. 113: 268-272 (1983).
7. G. F. Hurley, J. D. Fowler, and D. L. Rohr, Low dose cryogenic neutron irradiation effects in G-10CR, Cryogenics 23: 415-420 (1983).
8. H. W. Weber, E. Kubasta, W. Steiner, H. Benz, and K. Nylund, Low temperature neutron and gamma irradiation of glass

fiber reinforced epoxies, J. Nucl. Mater. 115: 11-15 (1983).
9. B. Korukonda, J. C. Conway, Jr., and R. A. Queeney, The effect of radiation and cryogenic temperature on the fatigue resistance of G-11CR glass-cloth/epoxy laminates, J. Nucl. Mater. 115: 197-202 (1983).
10. C. E. Klabunde and R. R. Coltman, Jr., Debonding of epoxy from glass in irradiated laminates, J. Nucl. Mater. 117: 345-350 (1983).
11. H. Yamaoka, K. Miyata, Y. Nakayama, and H. Yoshida, Mechanical properties of flake-reinforced plastics at low temperatures, in: "Proceedings of the International Cryogenic Materials Conference," Butterworth, Surrey (1982), pp. 282-285.

LOW TEMPERATURE PROPERTIES PROBED BY SELECTIVE LASER-EXCITATION

U. Bogner

Institut Physik III - Angewandte Physik
Universität Regensburg
Regensburg, F.R.G.

ABSTRACT

A sensitive probe for the analysis of low temperature properties of nonmetallic materials is provided by persistent spectral hole burning. The microscopic probe is based on narrow holes in the spectra of optical centers in solids, e.g. dye molecules in polymers or dyes adsorbed at crystal surfaces. The investigations concern thermal, electrical and mechanical properties of amorphous and crystalline materials. In the case of thermal properties the method is combined with heat-pulse technique and provides information about thermal conductivity and phonon processes in the double-well potentials of amorphous solids. Electric field effects are also studied, demonstrating applications in optical data storage.

INTRODUCTION

The selective laser-excitation of optical centers in nonmetallic materials, e.g. matrix-isolated dye molecules embedded in low concentrations in polymers, provides - in particular via the method of persistent spectral hole burning[1,2] - a sensitive probe for the analysis of the low temperature properties of solids.

The narrow spectral holes probe phonon processes[3-5], ultra sound[6] and electric field effects[7] which are connected with the physical origin of thermal, mechanical and electrical properties of the materials. The phonon processes concern the photophysical mechanism of the persistent spectral hole burning and the filling of the spectral hole by heat pulses. The results provide information about phonon processes not only in the material in which the dye molecule is embedded - as in the case of amorphous polymer films - but the results demonstrate also new methods of detecting high frequency phonons up to the terahertz range, which can be used for any material. This phonon detection is realized by embedding the dye molecules in Langmuir-Blodgett-films[8] i.e. ultrathin monomolecular layer systems, or it is realized by adsorption[9] of the dye molecule at the surface of the material. Persistent spectral hole burning and heat-pulse induced refilling is explained by the photophysical model[3] of matrix-shift variations caused by phonon-induced transitions in asymmetric double-well potentials[10,11]. The model was confirmed by direct proof of the matrix-shift variations[12].

Persistent spectral hole burning provides also applications e.g. in the field of optical data storage in the frequency domain[13,14] or even optical data storage in the electric field domain or voltage domain[8,15-18].

PERSISTENT SPECTRAL HOLE BURNING OF DYES AND HEAT-PULSE INDUCED REFILLING OF SPECTRAL HOLES

In Fig. 1 part of the energy level diagram of a dye molecule is shown to demonstrate the selective laser-excitation in the absorption range of the pure electronic transition and the different optical transitions which contribute to the low temperature fluorescence emission in the spectral range of the first vibronic transition. In this transition the lowest, intramolecular vibration is excited (see level 1 in the ground state S_0 of Fig. 1) which is coupling to the optical transition from the first excited singlet state S_1 to the ground state S_0. The insert above right in Fig. 1 demonstrates the density N(E) of occupation of the energy levels with energy E. The solid line in this insert shows the statistical distribution of the dye molecules because of different electronic transition energies of the dye molecules in their different sites in the amorphous matrix with different dye-matrix interaction. By selective laser-excitation the dye molecules with the energy level E_0 are resonantly excited in their pure electronic transition (0'←0), and therefore the emission spectrum does not show the large inhomogeneous line broadening but it contains narrow zero-phonon lines[2] of the pure electronic (0'→0) and the different vibronic transitions (0'→1), (0'→2) etc. The optical transitions in the spectral range of the first vibronic transition (0'→1) are depicted in Fig. 1. The vibronic relaxation of this

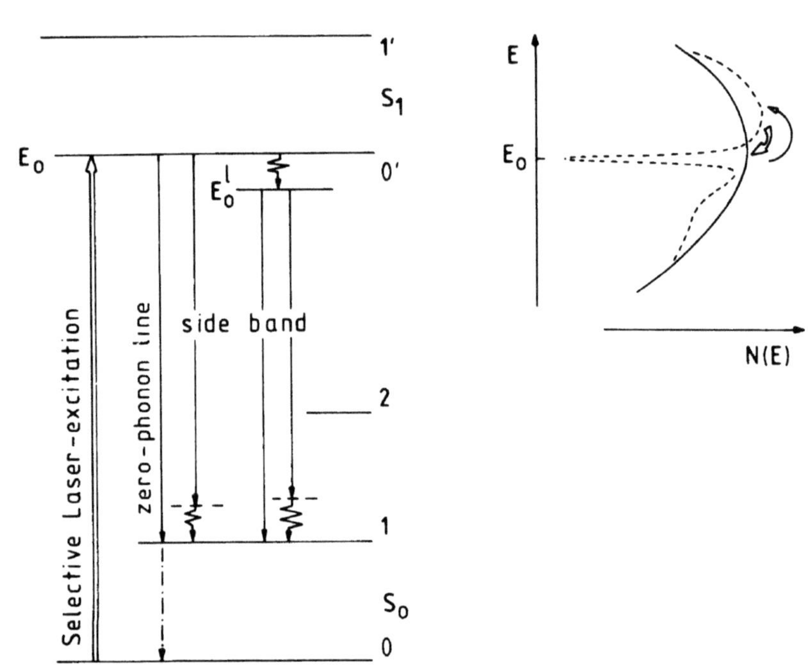

Fig. 1. Energy level diagram of selectively laser-excited dye molecules matrix-isolated in polymers. The insert above right demonstrates a schematic diagram showing the density N(E) of occupation of the levels with energy E.

level 1 is characterized by a dashed-dotted line. The zero-phonon line is due to an optical transition without involvement of a phonon. In the emission spectrum there are also broad sidebands beside the zero-phonon lines. These sidebands are due to optical transitions with simultaneous generation of phonons (see wavy lines in Fig. 1). The sidebands contain not only the phonon sidebands of the dye molecules, resonantly excited by the laser at the level E_0 in their pure electronic transition but also the zero-phonon lines and the phonon sidebands of lower lying levels E_0^1 absorbing the laser light in their phonon sideband in the first excited singlet state S_1.

In the course of laser excitation of dyes embedded in polymers a narrow hole is burnt at E_0 in the density N(E) of occupation of the levels (see dashed line in the insert of Fig. 1). The linewidth of these persistent spectral holes[1,2] is for almost all dyes substantially less than 1 cm^{-1} at the temperature of liquid helium and it can be measured by absorption spectroscopy or by fluorescence excitation with a narrow band laser. The reduction of the occupation of dye molecules in the center of the spectral hole can be determined by measuring the decrease of the fluorescence intensity of a zero-phonon line. Persistent spectral hole burning was obtained for a large number of different dyes like perylene, tetracene, 9-aminoacridine embedded in different types of polymers like epoxy-resin, polyurethane, polyvinylalcohol, polyvinylbutyral, polyvinylchloride,

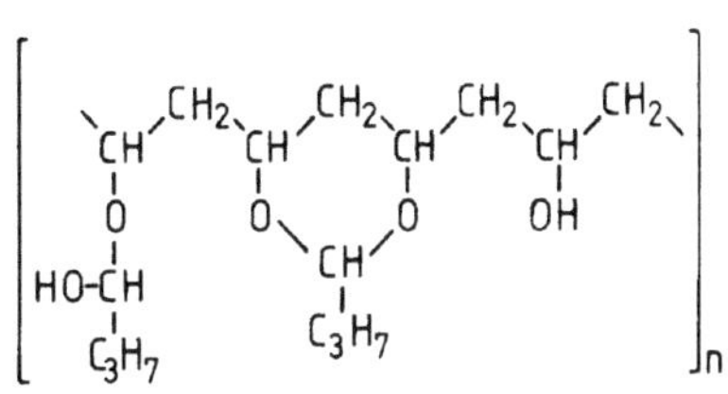

Fig. 2. Structural formula of polyvinylbutyral.

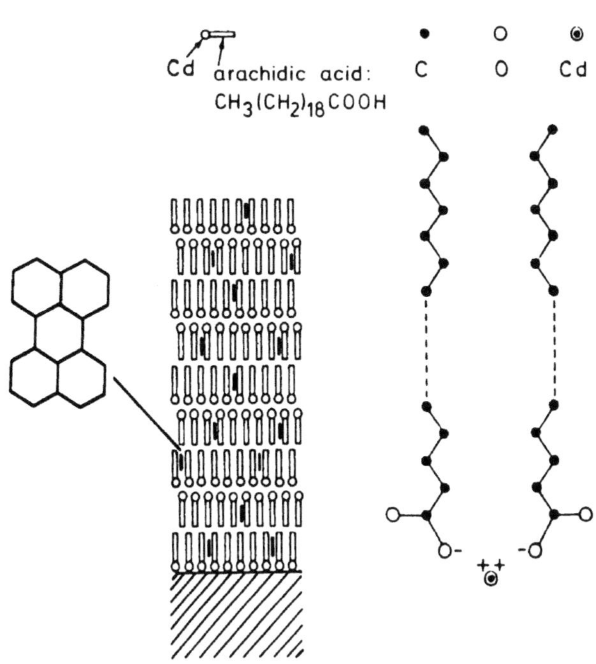

Fig. 3. Schematic drawing of the Langmuir-Blodgett-film.

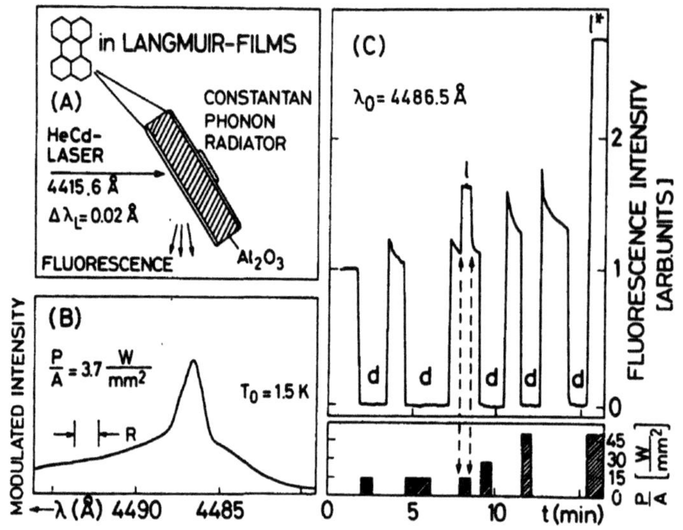

Fig. 4. Effects of heat pulses on the fluorescence of perylene in Langmuir-Blodgett-films (see text).[3]

polymethylmethacrylate and ethylcellulose[12]. Thin amorphous polymer films of high optical quality have been obtained using polyvinylbutyral. The structural formula of polyvinylbutyral (PVB) is shown in Fig. 2. PVB films doped with dyes could be prepared thinner than about 1 μ. A small thickness is necessary for the investigation of thermal conductivity and the basic physics of phonon processes in polymers. In the case when we apply our method of phonon detection with high temporal resolution to the study of thermal conductivity and phonon propagation or phonon scattering in different types of nonmetallic materials e.g. sapphire or quartz single crystals then ultrathin detector films prepared at the surface of the crystals are necessary. This is because in a noncrystalline detector film, like a polymer film, the mean free path of phonons in the terahertz range (10^{12} s^{-1}) is of the order of several 100 Å[19] and therefore e.g. the time resolution of the polymer film detector is strongly reduced. The problem of the film thickness could be overcome by using Langmuir-Blodgett-films[3,4,8]. A schematic drawing demonstrating a Langmuir-Blodgett-film of several monomolecular layers of Cadmium arachidate is shown in Fig. 3. Each layer is 28 Å thick. The aromatic dye molecule perylene was embedded in a low concentration (about 10^{-2} mol). In the first experiments[3] concerning the effects of phonons on the fluorescence of the selectively laser-excited perylene molecules

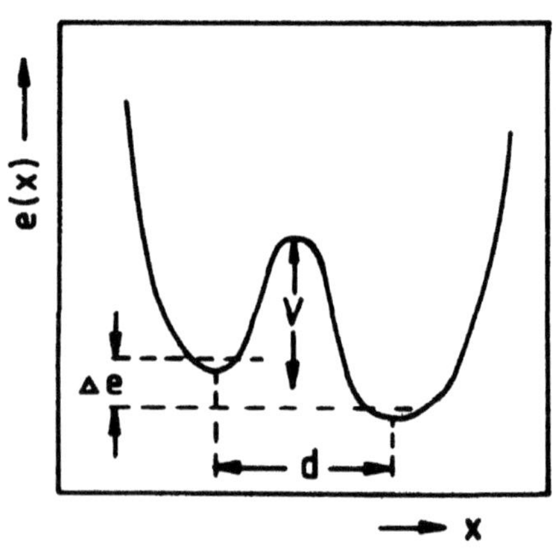

Fig. 5. Asymmetric double-well potential (see text).

this Langmuir-Blodgett-film was prepared at the surface of a sapphire plate. Persistent spectral hole burning was obtained by narrow-band excitation of the perylene molecules in the purely electronic transition with the blue line (wavelength: λ_L = 4415.6 Å) of a HeCd laser. By voltage pulses, applied to an evaporated constantan heater film, short heat pulses were generated in order to study the effects of phonons on the fluorescence of the perylene molecules in the spectral range of the first vibronic zero-phonon line. The power density P_H/A of the heat pulses was varied between 1 and 100 W/mm². The sample was immersed in superfluid He at T_0 = 1.5 K. The phonon-modulated fluorescence intensity demonstrated that there is a filling in the center of the spectral hole varying in the rhythm of the heat pulses. A new type of phonon detection i.e. a phonon memory[3] is shown in Fig. 4. After hole burning the fluorescence intensity of the first vibronic zero-phonon line (wavelength: λ = 4486.5 Å) is shown as a function of time t. On blocking the laser beam in the dark pauses d, heat pulses were irradiated. After the dark pause an increased fluorescence intensity was observed, decaying nonexponentially in some minutes to its original value. This provides a type of high spatial resolution photographic registration or erasable memory device for the detection of phonons up to the THz range[3,4].

The results are explained by the photophysical model[3] of persistent spectral hole burning and phonon-induced refilling of spectral holes. This model is based on matrix-shift variations caused by phonon-induced transitions in double-well potentials[10,11] with a large wide variety of barrier heights V and asymmetries Δe (see Fig. 5). These double-well potentials have been originally introduced in order to explain anomalous thermal and acoustical properties[20] of noncrystalline solids by the two-level systems[10,11,20] which are due to tunneling processes in these double-well potentials. In the photophysical model the matrix-shift-variations are caused by phonon-induced hopping processes in which the barrier of the double-well potentials is crossed. At low temperatures the hopping process over the barrier can be induced by nonthermal phonons generated by optical transitions of the dye molecules. The change in the environment of the dye molecule, connected with such a process, can alter the matrix-shift variation $\Delta\varepsilon$. The process involves a change of the level $E_0 = h\nu_L$, resonantly excited at the laser frequency ν_L, to a changed level $E_c = E_0 + \Delta\varepsilon$. The dyes in these E_c cannot be excited by the laser light, if an E_c is reached such that

$$E_c > h\{\nu_L^{max} + \Delta\nu_A(T)\} \quad (1)$$

where ν_L^{max} is the maximum frequency in the laser line and $\Delta\nu_A(T)$ is the thermal line broadening of the absorption profile at the sample temperature T. If this E_c is reached by a hopping process - and not by a tunneling process - the dye molecules cannot return to E_0 by themselves, since the phonon necessary for the reverse hopping process can no longer be generated by a laser photon at this dye site. By phonon irradiation, e.g. by a heat pulse, the number of dye sites at E_0, resonantly excited by the laser can be increased again. A transformation from E_c back to E_0 is possible for those dye sites for which the energy of phonons present at a given power density of the heater is high enough for a reverse hopping process.

The heat pulse experiments shown in Fig. 6 demonstrate that there is not only a phonon memory effect, providing a method for the detection of phonons after their irradiation but there is also a real-time phonon effect providing a method for the real-time detection of the phonons during their irradiation. In Fig. 6 the fluorescence in the range of the first vibronic zero-phonon line (0'→1) was measured

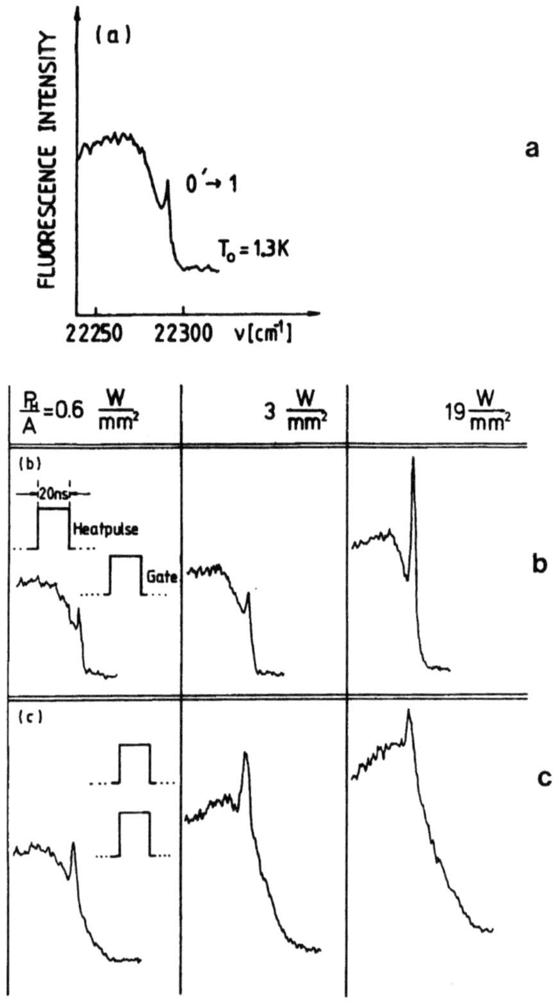

Fig. 6. Phonon memory and real-time effects of the phonons with high time resolution (see text).[5]

by a single-photon counting system with a gate (width: ≅ 20 ns) which could be delayed with respect to the voltage pulses (pulse width: 20 ns, repetition rate: 200 KHz) applied to the constantan heater. The upper three spectra in the tabular of Fig. 6 were recorded while the gate was shifted beside the heat pulse i.e. we measured the phonon memory which is caused by irreversible refilling of the spectral hole (see bended arrow in Fig. 7). The lower three spectra were recorded while the gate and the heat pulses were coincident, i.e. we measured the real-time effects of the heat pulses on the fluorescence which are due in particular to a reversible filling in the center of the spectral hole because of phonon-induced or thermal line broadening of the absorption profile (see dotted profile for the bath temperature T_0 and dashed profile for $T_1 > T_0$ in Fig. 7) of the dye

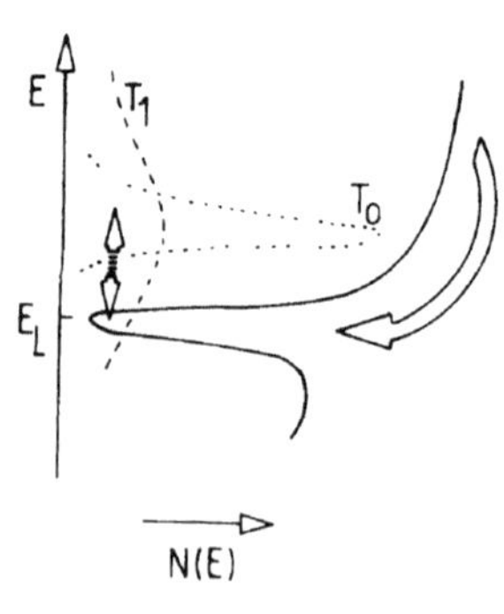

Fig. 7. Density of occupation, demonstrating the real-time phonon effects (see text).

molecules beside the spectral hole. In both cases the power density $P_H/A = 0.6\ W/mm^2$; $3\ W/mm^2$ and $19\ W/mm^2$, which correspond to calculated heater temperatures $T_H = 8$; 12; and 20 K respectively. The fluorescence in the anti-Stokes region of the zero-phonon lines in the lower spectra of Fig. 6 is mainly caused by phonon-assisted fluorescence excitation of dye molecules in higher lying levels $E_h > E_L$, which cannot absorb a laser photon without simultaneous absorption of a phonon.

In the case of the phonon memory we observed a logarithmic dependence of the refilling of the spectral hole on the number N of the irradiated heat pulses or to be precise there is a logarithmic dependence on the time of the phonon irradiation.[21] This result and the observation that there is a linear increase of refilling with increasing heater temperature is explained by a constant density of states of the double-well potentials concerning their barrier heights in an energy range which corresponds to phonon frequencies up to several THz.

We have observed persistent spectral hole burning and phonon-induced refilling also with dye molecules adsorbed at the surface, e.g. of a sapphire single crystal or at the surface of the microcrystals in polycrystalline paraffins or in alumina powder[9,12]. Concerning the physical nature of the double-well potentials the results suggest a generalized microscopic model in which double-well potentials are formed by adsorbed molecules which were bound by weak intermolecular forces to surfaces or interfaces e.g. at the boundary of the small free volume regions present in all disordered materials.

If the adsorbed perylene molecules or Langmuir-Blodgett-films doped with perylene molecules are prepared at the surface opposite to the surface containing the constantan heater film, then thermal conductivity and in particular the ballistic propagation of phonons with sound velocity or phonon scattering can be investigated by time of flight measurements with high temporal resolution, provided by the real-time phonon effects described above. By this procedure we investigated phonon propagation in sapphire and quartz single crystals and in polymer films. With sapphire and quartz single crystals we observed ballistic propagation of heat pulses with the velocity of the transverse and longitudinal acoustic sound waves.[21] In quartz single crystals we investigated also phonon focussing and in the polymer films the results indicate strong inelastic phonon scattering.

ELECTRIC FIELD EFFECTS AND POSSIBLE APPLICATIONS

Refilling in the center of the narrow spectral holes can also be obtained by applying an electric field to the sample. This process can be used not only as a sensitive probe to study electric field effects in polymers but it can be used also for possible applications in particular in optical data storage. The sample configuration used in these investigations is shown in Fig. 8. The sample consists of a thin film of polyvinylbutyral doped with dye molecules (e.g. perylene or 9-aminoacridine) in a low concentration ($\sim 5 - 10^{-4}$ mol). It is coated with two electrically conducting electrodes one of which is transparent to light, by using e.g. indium tin oxide.

Fig. 9 shows the fluorescence in the spectral range of the first vibronic ($0' \rightarrow 1$) transition excited by a HeCd laser. Curve 1 and 2 are recorded before and after hole burning respectively, with no external electric field. Curve 3 was obtained with an electric field strength

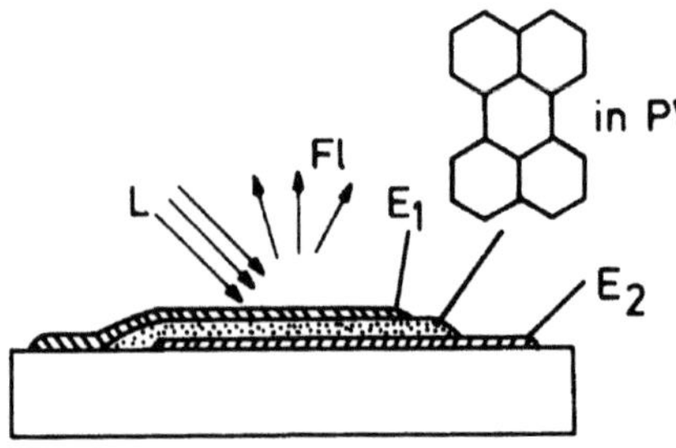

Fig. 8. Sample configuration for the electric field effect.

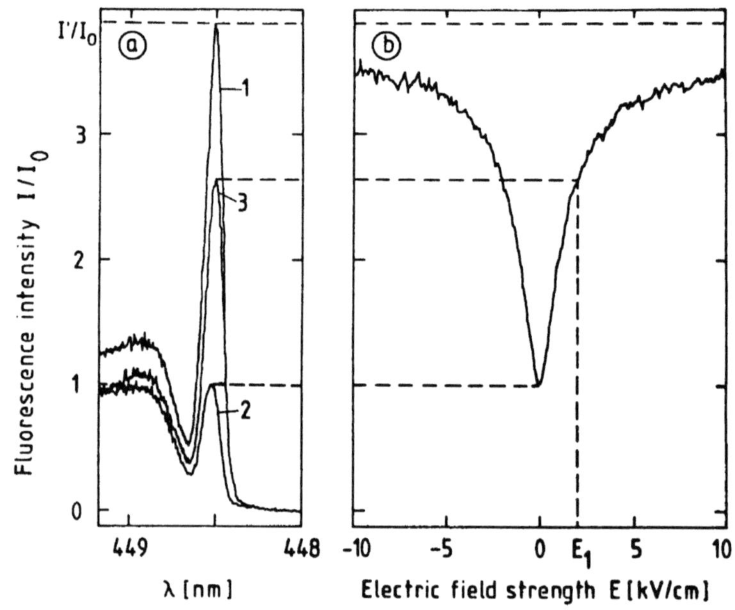

Fig. 9. Electric-field effects on the fluorescence spectrum (see text).[7]

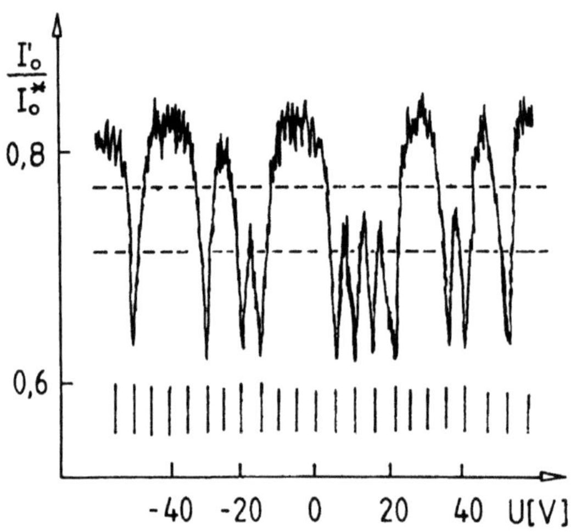

Fig. 10. Stable multiple holes in the fluorescence intensity versus voltage diagram, demonstrating optical data storage.[17]

of 2.0 kV/cm. The observed fluorescence increase is due to the electric-field induced filling in the center of the spectral hole. Fig. 9 b shows the normalized fluorescence intensity I/I_0 of the zero-phonon line in dependence of the electric field strength F. This diagram was fitted in a calculation[7] based on a linear dependence of the level shifts $\Delta\nu_i$ on the electric field change ΔF

$$\Delta\nu_i = \beta_i \Delta F \qquad (2)$$

The factor β_i is ascribed to the interaction of the dye molecule with the asymmetric amorphous matrix.[7] The electric field-induced filling in the center of the spectral hole has been applied to determine the linear electron-phonon interaction which is characterizing the low temperature properties of the dye-matrix system.[16]

A technical application of the electric-field effects is demonstrated in Fig. 10. In this diagram stable multiple holes are burned at various voltages. These multiple holes in the fluorescence versus voltage diagram of 9-aminoacridine in polyvinylbutyral demonstrate possible applications for optical data storage or to be precise for a voltage-tunable optical data storage in which the electric field domain is used as a storage dimension in addition to the frequency domain.

ACKNOWLEDGEMENTS

The author wishes to thank Professor M. Maier for valuable discussions and to Hoechst Company (Frankfurt) for the polyvinylbutyral material. Financial support by the Deutsche Forschungsgemeinschaft is gratefully acknowledged.

REFERENCES

1. L. A. Rebane, A. A. Gorokhovskii and J. V. Kikas, Appl. Phys. B 29: 235 - 250 (1982).
2. R. I. Personov in: "Spectroscopy and Excitation Dynamics of Condensed Molecular Systems", V. M. Agranovich and R. M. Hochstrasser, eds., North-Holland, Amsterdam (1983), pp. 555 - 621.
3. U. Bogner, Phys. Rev. Lett. 37: 909 - 912 (1976).
4. U. Bogner and G. Röska, J. of Luminescence 24/25: 683 - 686 (1981).
5. U. Bogner and G. Röska, in: "Phonon Scattering in Condensed Matter", W. Eisenmenger, K. Laßmann, and S. Döttinger, eds., Springer-Verlag, Berlin (1984), pp. 395 - 398.
6. H. Lengfellner, T. R. Gosnell, R. Tkachand, A. Sievers, Appl. Phys. Lett. 43: 437 - 439 (1983).
7. U. Bogner, P. Schätz, R. Seel and M. Maier, Chem. Phys. Lett. 102: 267 - 271 (1983), and references therein.
8. U. Bogner, G. Röska and F. Graf, Thin Solid Films 99: 257 - 261 (1983).
9. U. Bogner and G. Röska in: "Surface Studies with Lasers", F. R. Aussenegg, A. Leitner, M. E. Lippitsch, eds., Springer-Verlag, Berlin (1983), pp. 126.
10. P. W. Anderson, B. J. Halperin and C. M. Varma, Philos. Mag. 25: 1 (1972).
11. W. A. Phillips, J. Low. Temp. Phys. 7: 351 (1972).
12. U. Bogner and R. Schwarz, Phys. Rev. B 24: 2846 - 2849 (1981).
13. D. Haarer, H. Wolfrum, J. Friedrich and A. R. Gutierrez IBM J. Res. Develop. 26: No. 2, 198 (1982).
14. G. Björklund and G. Castro, IBM Research Report RJ 3287 (39813) San Jose, Calif., USA (1981).
15. U. Bogner, R. Seel and F. Graf, Appl. Phys. B 29: 152 (1983).
16. U. Bogner, P. Schätz, K. Beck and M. Maier, Chem. Phys. Lett. 110: 528 - 532 (1984).
17. U. Bogner, P. Schätz, K. Beck and M. Maier, Electric-Field-Induced changes of Persistent Spectral Holes in Amorphous Solids and their Applications, in: "Digest for the Int. Conference on Lu-

minescence", ed. by Optical Society of America (1984), pp. WE 13.
18. U. Bogner, K. Beck and M. Maier, Electric Field Selective Optical Data Storage Using Persistent Spectral Hole Burning, Appl. Phys. Lett. 46: 534 - 537 (1985).
19. R. C. Zeller and R. O. Pohl, Phys. Rev. B 4: 2079 (1971).
20. "Amorphous Solids - Low Temperature Properties", W. A. Phillips, ed., Springer-Verlag, Berlin (1981).
21. U. Bogner, G. Röska and P. Schätz, Phonon Processes in Disordered Systems, Probed by Spectral Hole Burning and Refilling of Dyes, in: "Digest for the Int. Conference on Luminescence", ed. by Optical Society of America (1984), pp. WE 12.

OPTICAL INVESTIGATION OF PRESSURE-INDUCED MATRIX CHANGES IN POLYMERS UNDER LOW TEMEPRATURE CONDITIONS

W. Richter, G. Schulte and D. Haarer

Physikalisches Institut
der Universität Bayreuth
Postfach 3008, D-8580 Bayreuth, F.R.G.

ABSTRACT

Photochemical hole burning (PHB) is a laser spectroscopy method yielding high optical resolution at temperatures below 30 K. The method utilizes dye molecules embedded in polymer hosts to detect minute matrix changes resulting in optical changes of linewidths and line positions of as little as 10^{-3} Å. Using the dye molecule phthalocyanine in the polymer matrices polyethylene (PE) and polymethylmethacrylate (PMMA), it could be shown, that pressure changes of as little as 100 hPa lead to appreciable broadening of the line profile. At higher pressures up to $3 \cdot 10^4$ hPa we observed irreversible line broadening phenomena, which we attribute to irreversible site changes of the host-guest system. Temperature-induced line broadening will be compared to the pressure data.

INTRODUCTION

The technique of photochemical hole burning (PHB) provides a convenient experimental method for extending the optical spectral resolution well below the inhomogeneous bandwidth[1-3]. It is a special kind of saturation spectroscopy, where, a narrow bandwidth laser excitation is used to burn very sharp and stable photochemical holes into the absorption bands of guest molecules. Since the optical linewidth of a typical PHB experiment can be as small as the homogeneous linewidth of the dopant molecule, the gain in resolution ($\gamma/\Delta\gamma$) can, at low temperatures, be on the order of 10^3 to 10^5.

The homogeneous linewidth of absorbers embedded in amorphous hosts, such as inorganic glasses[4], organic glasses[5] and polymer hosts[6,7] are always broader than in crystalline environments. Various theoretical models have been proposed to explain these fast dephasing processes[8-10]. All of the theories involve the interaction of the local absorber with a distribution of two level systems (TLS) or tunneling states, which are characteristic of the amorphous host. The absolute value of the holewidth of a given absorber molecule depends on the host system, respectively, and is very sensitive against perturbations of the system by external parameters, giving rise to line shifts and broadenings.

In the following, we would like to report first experimental data on PHB spectra of doped polymer samples under uniaxial pressure. The influence of external pressure on the absorption and emission spectra of aromatic molecules embedded in polymer hosts has, to our knowledge, thus far only been investigated with straight forward spectroscopic methods (for a review see ref. 11). Typical experiments with aromatic molecules in PMMA matrices show a pressure induced red shift of the inhomogeneous band on the order of $\Delta\nu/\Delta p = 0.7\cdot10^{-2} - 3\cdot10^{-2}\ cm^{-1}/10^3$ hPa. The available experimental data[11] focus on the pressure regime $\geq 10^6$ hPa since, at smaller pressures, the shifts of the broadband absorption or emission spectra cannot be measured easily.

The high resolution method of PHB allows very sensitive experiments in the lower pressure regime down to 100 hPa. The investigation of reversible and irreversible line broadening gives some insight in the aspect of low temperature, pressure-induced matrix changes of polymers and into the distribution of barrier heights V_o of the TLS.

EXPERIMENTAL AND RESULTS

Our PHB systems are polyethylene (PE) and polymethylmethacrylate (PMMA) as the host matrices and the free base molecule phthalocyanine (H_2Pc) as the dopant. The low temperature photochemistry, which leads to the hole burning in the H_2Pc system, has been attributed to a tautomerization of the two central protons of the dye molecule[7]. The investigated transition is the zero-phonon component of the $S_1 \leftarrow S_o$ transition of the H_2Pc molecule. Fig. 1 shows the gross feature of the photochemical change of the inhomogeneous broadened $S_1 \leftarrow S_o$ absorption band of H_2Pc upon laser irradiation at a wavelength λ_L. Typical concentrations of the dopant are 10^{-4} to 10^{-5} mole/liter. As can be seen from Fig. 1 the photoproduct is characterized by a 'photochromic shift' on the order of 15 nm to longer wavelengths. It should be noted, that the width of the zero-phonon line (peak at λ_L) is considerably broadened in this figure due to the long laser irradiation[12] which was necessary to photoconvert a sizeable fraction of the dye molecules and to show clearly the spectral localization of the photoproduct. In the following experimental situations the laser irradiation was limited, so that the corresponding hole depth was below 30% of the intensity at maximum absorption; this way line broadening phenomena were minimized. Also the

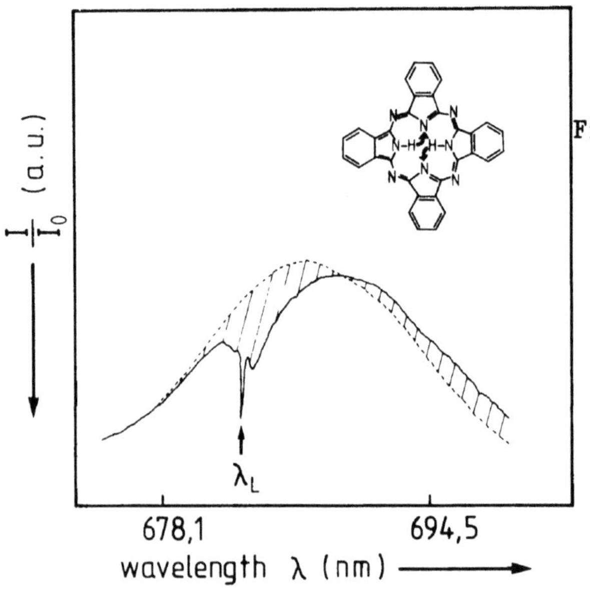

Fig. 1. Inhomogeneous absorption band of the lowest electronic transition of the H_2Pc molecule in PMMA
a) without laserphotochemistry (dashed line)
b) with a hole close to the photochemical saturation limit (full line)
Insert: Molecular model of the involved photochemistry

phonon sideband on the low energy side of the spectrum, which is usually referred to as the 'pseudo phonon wing' will not be further discussed. Its intensity is, at low irradiation dose, negligible[12].

Burning and probing of the holes were carried out with a tunable single frequency dye laser with a spectral resolution of about 1 MHz. Typical optical holewidths were, at 1.5 K, $6 \cdot 10^{-3}$ cm^{-1} in PE and $26 \cdot 10^{-3}$ cm^{-1} in PMMA. The holes are, at low temperatures (<10 K), permanent and show no measurable changes over several hours (i.e. typical experimental times).

The PE and PMMA samples were polished to flat discs and then exposed to uniaxial pressure between quartz discs. The sample holder was immersed in a liquid He-cryostat, whose temperature was kept at 1.5 K. The experimental range of the applied pressure could be varied from 10^2 to $3 \cdot 10^4$ hPa. The insert of Fig. 3 shows the experimental sample configuration. The experimental results show, that the PHB method is extremely sensitive to minute spectral changes at low pressures. Therefore we did, in most cases, burn the holes after applying a 'bias pressure' or 'pre-pressure'. Subsequently we investigated the width and the position of the hole by changing the pressure in small steps to higher values. The experimental results show, that the 'pre-pressure' has, up to values of $3 \cdot 10^4$ hPa (our present experimental limit) no influence on the width of the 'test hole' i.e. the first hole, which is being burnt and measured prior to any pressure changes. The width corresponds to values, which have been reported earlier[3].

Fig. 2 shows the width and position of a photochemical hole burnt at a pre-pressure of $25.6 \cdot 10^3$ hPa and measured up to pressures of about $28.7 \cdot 10^3$ hPa. One can see from the figure, that within a pressure range

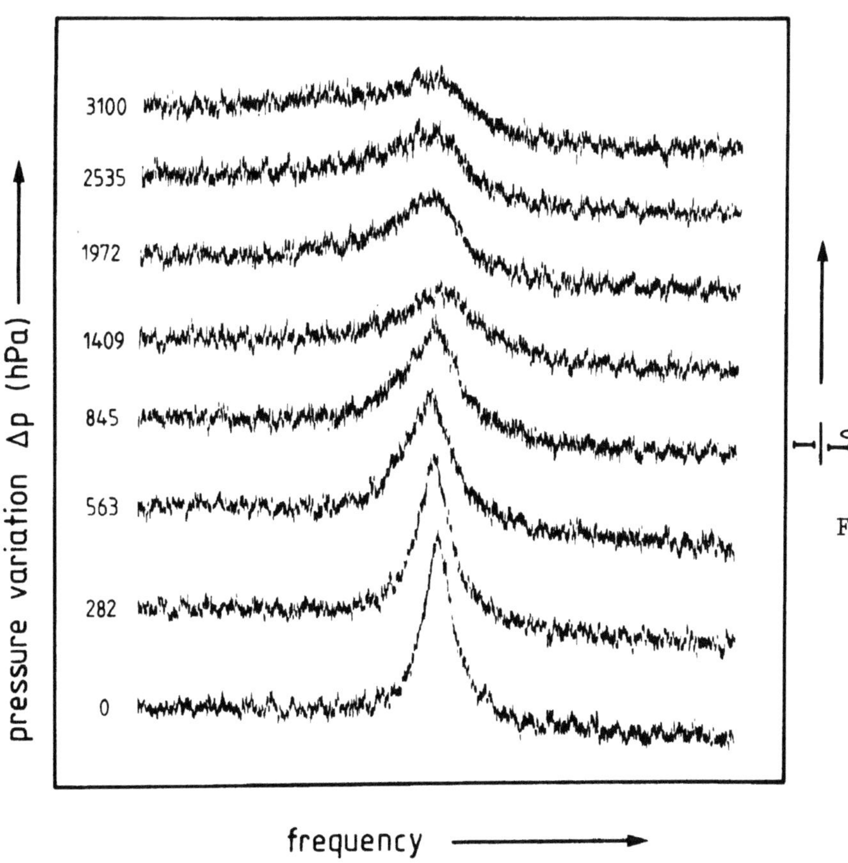

Fig. 2. Evolution of a hole specturm in PMMA (λ_L = 686 nm) measured at seven difference pressures Δp. Value of the prepressure $p = 26.5 \cdot 10^3$ hPa.

of about $3 \cdot 10^3$ hPa the hole shows an asymmetric broadening accompanied by a small shift to longer wavelengths. The hole area remains, within experimental error, unchanged.

The experimental data of the optical linewidths as a function of pressure have to be separated into two pressure regimes. In the regime up to pressures of $\Delta p \simeq 500$ hPa and the regime of $\Delta p \geqslant 500$ hPa. In the low pressure regime the change in linewidth is predominantly reversible; at higher pressure changes of Δp, irreversible effects have to be taken into account (see below). Fig. 3 shows the change of the hole width of H_2Pc in PE at a pre-pressure of $2.5 \cdot 10^3$ hPa. The increase in linewidth seems to follow a power law of

$$\Delta\Gamma \simeq (\Delta p)^\alpha \quad (1)$$

giving a straight line in a log-log plot.

Experiments to higher pressure changes Δp are shown in Fig. 4 for PE and PMMA. The lines shown in the figures connect the data points for five different values of the pre-pressure. It should be noted that the error bars are, for small linewidth changes, rather large on the logarithmic scales of Fig. 4. It should also be noted, that at high values of Δp (>500 hPa) the measured line changes show a pressure dependence, which is in some cases extremely nonmonotonic. We attribute this behavior to irreversible processes, which take place in the host matrix.

Hole broadening can also be observed by changing the sample temperature[13]. Fig. 5 shows the increase of the optical holewidth as a function of the temperature in the PE matrix. The four temperature

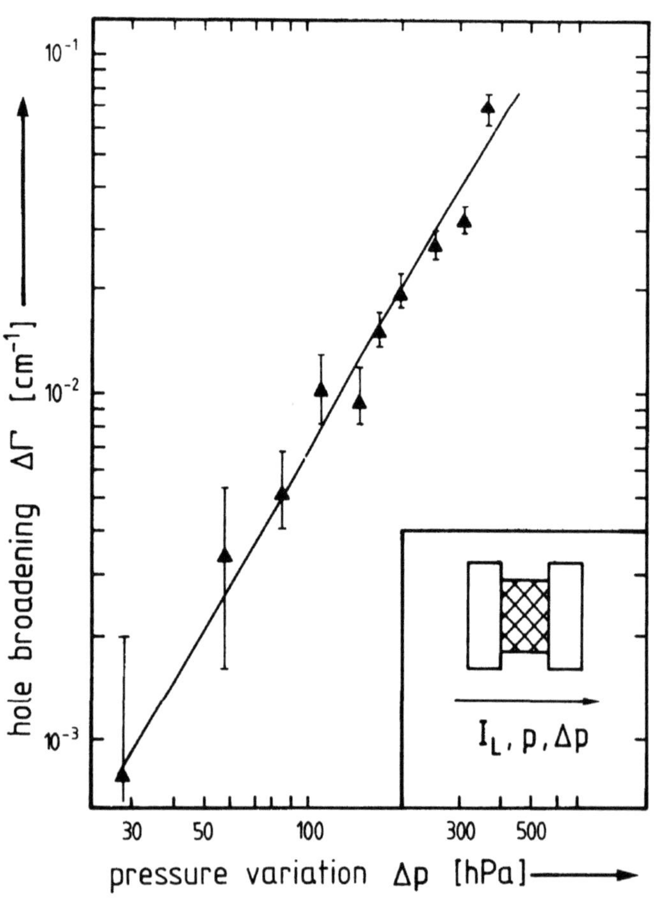

Fig. 3.
Broadening of the optical halfwidth of a photochemical hole in PE as a function of the pressure variation Δp (log-log plot). Pre-pressure value $p = 2.5 \cdot 10^3$ hPa.

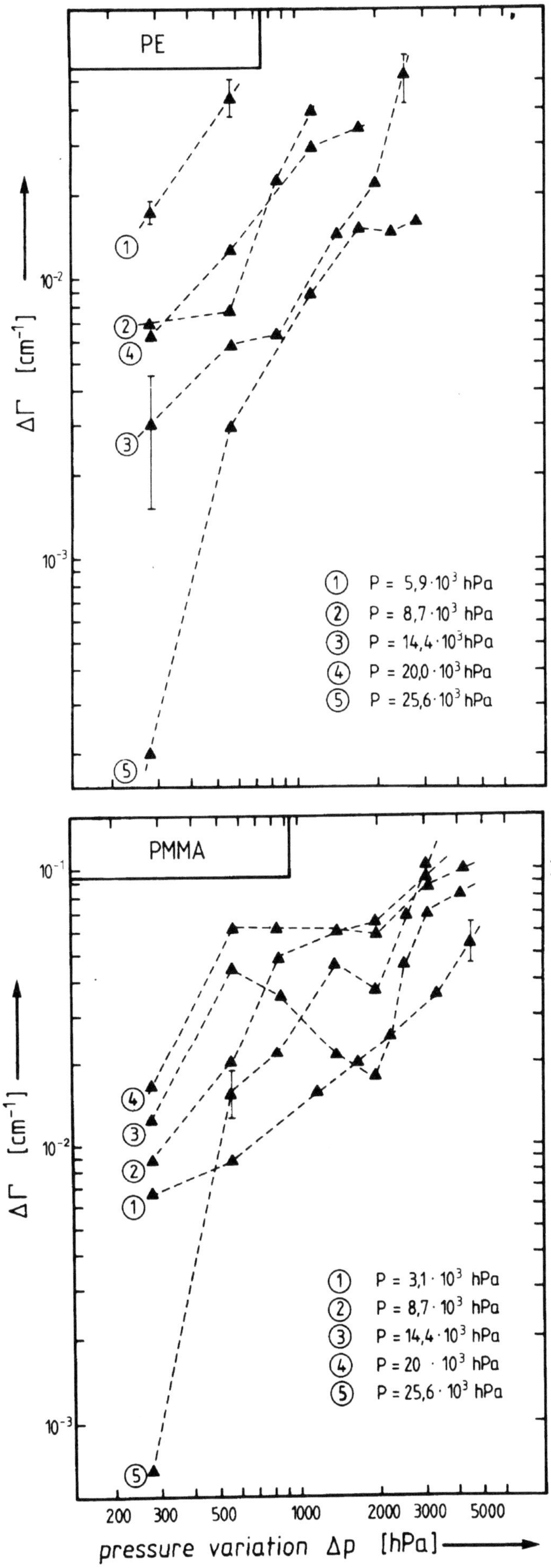

Fig. 4. Broadening of the optical halfwidth of different photochemical holes in PE (upper figure) and PMMA (lower figure) as a function of Δp. The encircled numbers correspond to the values of the prepressure.

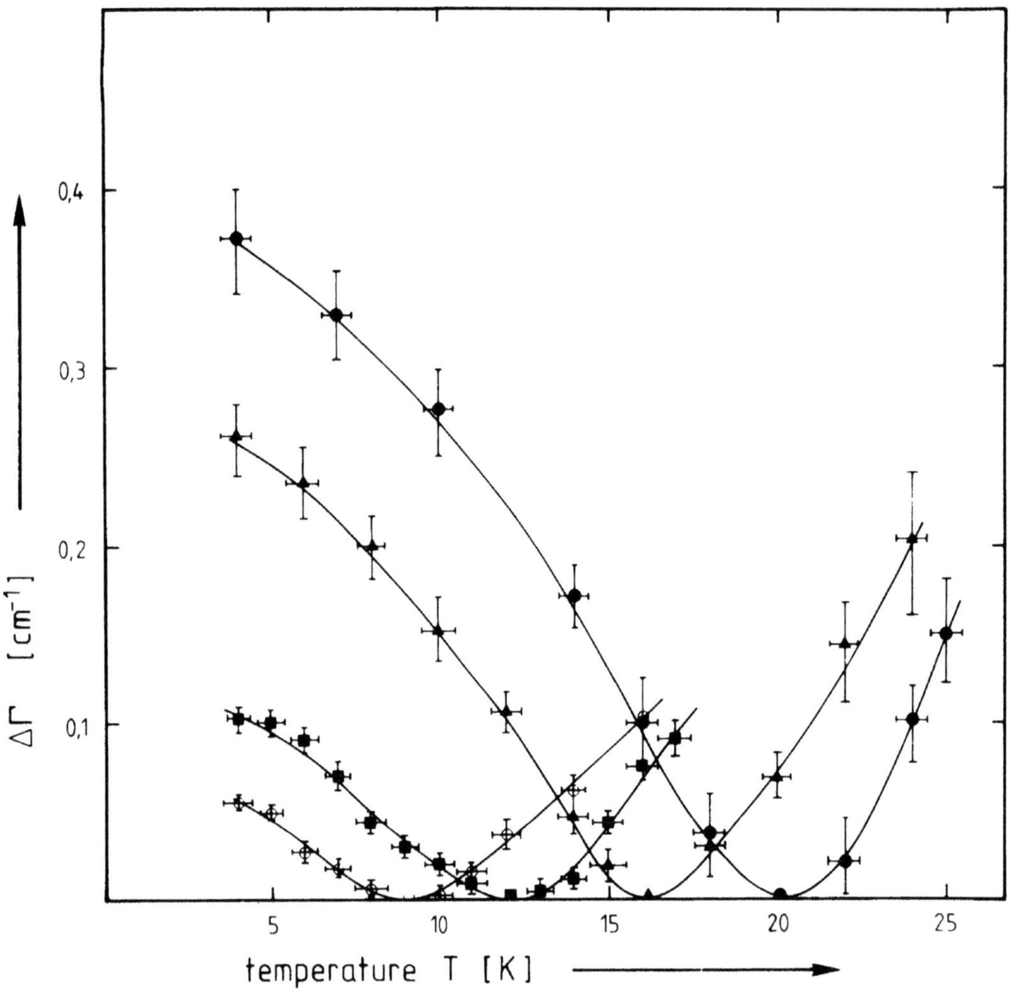

Fig. 5. Broadening of the halfwidth of a photochemical hole as a function of temperature in PE. The four temperature cycles belong to testholes burnt at 9K,12K,16K and 20K.

cycles correspond to four different holes (i.e. 'testholes') burned at four different temperatures. The amount of broadening to lower temperatures depends on the starting point of each cycle, that is the burning temperature of each testhole. This broadening measured at a fixed cycle temperature will be related to the pressure-induced hole broadening via internal shear strains, which are produced by both mechanisms. The temperature-induced broadening to lower temperatures is fully reversible. The different line broadening to higher temperatures will be discussed elsewhere[13].

DISCUSSION

The absorption line shape $A(\omega)$ of a photochemical zero-phonon hole, burnt at the frequency ω_L, can be described by a convolution of the line shape function $g(\omega)$ of an individual molecular absorber with the distribution $N(\omega)$ of photochemically converted molecules[3,12].

$$A(\omega) = C \int N(\omega_L - \omega')g(\omega - \omega')d\omega' \qquad (2)$$

C is a constant which contains the various molecular and experimental parameters (see Ref. 12).

The two different functions $g(\omega)$ and $N(\omega)$ of Equ.(2) can be independently investigated by appropriate experiments. The absorption profile $g(\omega)$ reflects the molecular T_1 and T_2 relaxation processes and is, within the experimentally investigated pressure range, independent of the applied pressure. Therefore our photochemical holes are, at the onset of each experiment, independent of the 'pre-pressure'.

With $g(\omega)$ being independent of the applied pressure, our experimental variation of the pressure change Δp after the hole burning experiment allows us to study the $N(\omega)$-function as a function of the applied pressure change. $N(\omega)$ corresponds to the frequency distribution of photochemically converted sites and can be looked upon as the 'site memory function' which is sensitive to reversible and irreversible frequency changes of the photochemically labeled sites. We refer to these changes, which lead to the observed overall change in lineshape, as to pressure induced spectral diffusion processes. The underlying mechanism is a redistribution of site energies of the optical centers, which form the envelope of the photochemical hole (i.e. centers, which have not been affected by the laser photochemistry). The experimental observation that the hole area remains constant during the Δp experiments proves, that the photoreaction is not affected or reversed by the pressure.

If we assume, that a polymer glass has volume elements of different densities, then one can expect, that the various sites, which originally absorb at one single frequency ω_i are shifted to different frequencies $\omega_i \pm \Delta\omega$. This is symbolically shown in Fig. 6 for three sites $a_i - c_i$, each consisting of three absorber groups contributing to the respective frequency ω_i. Since different absorber groups in environments of different 'local density' are expected to exhibit different pressure shifts (see upper graph in Fig. 6), the optical line, which is the envelope of the molecular absorbers which are unperturbed by the photochemistry at ω_L, spreads out over a frequency interval $\Delta\omega$. This frequency spread is due to the pressure shift of the individual absorbers. The asymmetric broadening of the line to higher difference pressures can be explained in terms of different red shifts of the individual absorber groups on a qualitative level (not shown in the figure).

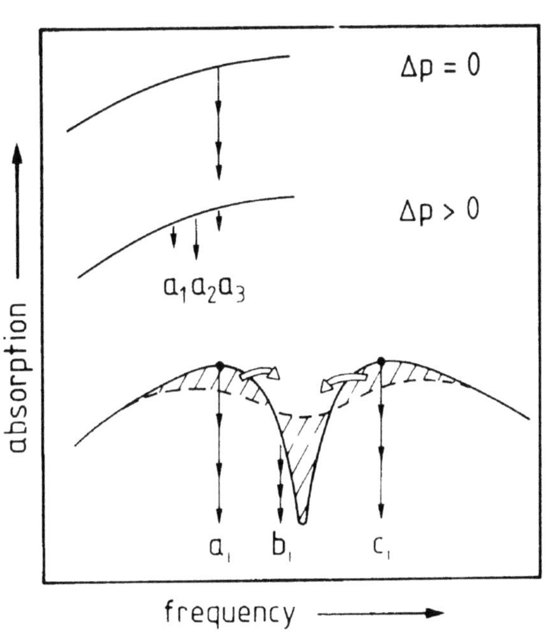

Fig. 6. Scheme for pressure-induced hole broadening. The figure shows the individual site shifts of the molecular absorbers a_i-c_i. The mechanism for the braodening of each individual site group is symbolically depicted in the upper graph.

A feature which we have observed experimentally, yet not been able to put into a quantitative framework, is the fact, that in the lower difference pressure regime the linewidth changes are accompanied by a strong reversible part after recycling to the start pressure value. Assuming a simple double potential well model for the glassy state[8-10], it might well be, that reversible pressure changes do not lead to changes in the TLS minimum energy, whereas other pressure changes lead to irreversible changes of the TLS minimum energies of the molecular ground states.

Experimentally the asymmetric broadening yields a frequency shift of $d\nu/dp \simeq 10^{-2} cm^{-1}/10^3$ hPa, which is in accordance with the values measured for the broadband shifts of aromatic molecules[11].

The experimentally observed increase in linewidth $\Delta\Gamma$ follows in the low pressure regime (Fig. 3) the general law of

$$\Delta\Gamma/\Gamma \simeq (\Delta p/p)^{\alpha} \qquad (3)$$

with α ranging between 1.5 and 2.5.

The fact, that we observe, at higher pressures, rather nonmonotonic increases of $\Delta\Gamma$ versus Δp is taken as being indicative of irreversible matrix rearrangements. If one again takes the TLS-model of the glassy state and if one assumes a wide range of barrier heights V_o, then one would expect irreversible effects to occur at low temperatures for systems with low barrier heights. The experimental observation of irreversible pressure effects above 500 hPa can be taken as evidence for the fact, that irreversible phenomena occur in the glassy state at values of pressure changes Δp at which a crystalline lattice would be considered as being completely dominated by elastic and, hence, reversible processes.

Because of the amorphous nature of the polymer matrices and the therefore disarranged hydrocarbon chains, the dye molecules will be subject to density fluctuations of the matrix. Volume changes ΔV induced by temperature or pressure variation will therefore lead to different shear strains throughout the sample. An explanation of the temperature data in Fig. 5 will take into account an increasing shear strain during lowering the matrix temperature[13]. The hole broadening $\Delta\Gamma$ of each temperature cycle can be well fitted within this model by

$$\Delta\Gamma \simeq T_B^2 - T_C^2 \qquad (4)$$

T_C and T_B are cycle and burning temperature respectively.

In the lower difference pressure region our hole broadening data are very similar to the temperature-induced broadening with respect to the reversible behavior and the absolut hole broadening values. The pressure-induced site changes of the dye molecules might therefore be based on the same underlying mechanism.

CONCLUSION

We have documented that PHB experiments can be utilized to extend the spectroscopy of glasses under uniaxial pressure to very low pressure regimes. The measured line shifts are in accordance with the shifts observed for broadband systems. The most remarkable findings of our experiments are effects of irreversible spectral diffusion, which occur at pressure changes of as little as 500 hPa. We attribute these changes to matrix rearrangements in the glassy state, which can, for the first time be measured due to the high resolution of the experiment and the 'memory' of the photochemically produced N(ω) distribution. This distribution registers spectroscopic site shifts which occur after the site selective photochemistry. The stable irreversible rearrangement of the dye molecules labelled via photochemistry leads to an interesting application[14] in the field of optical data storage in amorphous systems using the PHB method. In addition, in the field of the science of glassy materials at low temperatures, this method might be applicable for testing such materials with respect to matrix changes under external stress at a very early state.

ACKNOWLEDGEMENT

We thank our colleague J. Friedrich for many helpful discussions and we are grateful to the BASF-corporation for providing us with samples.

REFERENCES

1. B. M. Kharlamov, R.I. Personov and L.A. Bykovskaya, Opt. Comm. 12:191-193 (1974).
2. A. A. Gorokhovskii, R.K. Kaarli and L.A. Rebane, J.E.T.P. Lett. 20:216-218 (1974)
3. J. Friedrich and D. Haarer, Ang. Chemie, Int. Ed. 23:113-140 (1984)
4. J. Hegarty and W. M. Yen, Phys. Rev. Lett. 43:1126-1130 (1979)
5. G. Small in: "Spectroscopy and Excitation Dynamics of Condensed Molecular Systems", V. W. Agranovich and R. M. Hochstrasser, eds., Amsterdam, N.Y., Oxford (1983) pp. 515-554
6. R. J. A. P. Marchetti, M. Scozzafava and R. H. Young, Chem. Phys. Lett. 51:424-426 (1977)
7. A. R. Gutierrez, J Friedrich, D. Haarer and H. Wolfrum, IBM J. Res. Develop. 26:198-208 (1982)
8. J. M. Hayes and G. J. Small, Chem. Phys. 27:151-157 (1978)
9. S. K. Lyo and R. Orbach, Phys. Rev. B22:4223-4225 (1980)
10. P. Reineker, H. Morawitz and K. Kassner, Phys. Rev. B29:4546-4561 (1984)
11. H. W. Offen, Absorption and luminescence of aromatic molecules at high pressures, in: "Organic Molecular Photophysics", Vol. 1, J. B. Birks, ed., London, N.Y., Toronto (1973), pp. 103-151
12. J. Friedrich, J.D. Swalen and D. Haarer, J. Chem.Phys. 73:705-711 (1980)
13. G. Schulte, Dissertation thesis, Bayreuth (to be published)
14. W. Richter, "Löschverfahren für optische Datenspeicher", F.R.G.-patent pending

LOW FREQUENCY MEASUREMENTS ON POLYMETHYLMETHACRYLATE

N. Geis, G. Kasper and S. Hunklinger

Institut für Angewandte Physik II
der Universität Heidelberg
6900 Heidelberg, FRG

ABSTRACT

Variations of the sound velocity and the internal friction have been measured in commercial polymethylmethacrylate from 0.3 K to 290 K at frequencies below 1 kHz using a vibrating reed technique. Below 6 K the polymer shows the acoustic properties known from inorganic glasses: a temperature independent plateau in the absorption and a logarithmic variation of the sound velocity with temperature. The results at low temperatures are discussed within the framework of the Tunneling Model of glasses.

INTRODUCTION

At very low temperature the random structure of amorphous polymers is frozen in. Warming up allows certain molecular units of different size to move. This mobility enables those structural units to rearrange or in other words to relax, after external perturbation. Therefore the onset of the molecular motion of specific units is readily seen as loss peaks due to structural relaxation in mechanical measurements, or dielectric experiments if the units involved are polar. Going to higher and higher temperatures even parts of the main chain itself become mobile and give rise to strong losses close to the glass transition temperature. For the sake of simplicity the motions of different parts of the molecules are often assumed to be independent, but in real polymers the various processes frequently overlap. Thus a clean separation of different molecular motions very often becomes extremely difficult.

Below room temperature pure polymethylmethacrylate (PMMA) does not show pronounced loss peaks as compared to other polymers, the damping decreases gradually with decreasing temperature[1]. Slightly above room temperature a relaxation peak is found in both dielectric and mechanical studies[2]. It is caused by the re-orientational motion of the entire side chain. This rotational motion is hindered by the presence of potential barriers due to adjacent backbone methyl groups. Close to the glass transition temperature at about 380 K another peak is observed which is common to all polymers. It originates from large-scale translational and rotational motions of the main chain.

In our investigation we concentrate on the elastic behaviour of

PMMA below 10 K. At these temperatures an indication of a loss maximum has been found[1]. So far it was thought that the low temperature properties are primarily determined by the re-orientational motion of pendant methyl side chains. We suggest that the observed properties can be explained in the same way as the so-called cryogenic anomalies observed in almost all glassy solids[3]. These anomalies arise from low-energy excitations present in amorphous solids in addition to "ordinary vibrational states" known from crystals. The existence of these additional states in PMMA has clearly been demonstrated by measurements of the heat capacity extending to temperatures well below 1 K. In contrast to what is known in pure dielectric crystals, the specific heat is not proportional to T^3 but rather linear in temperature. Studies of the thermal conductivity support this interpretation. The mean free path of thermal phonons is not limited by geometrical boundary scattering even below 0.1 K, but by intrinsic scattering processes. Consequently thermal conductivity is not proportional to T^3 as in crystals at low temperatures but approximately proportional to T^2 as in inorganic glasses.

So far ultrasonic[4] and Brillouin scattering[5] experiments have been valuable tools to study the influence of these low-energy excitations on the elastic properties of PMMA. Recent measurements in inorganic glasses and metallic glasses[6] suggest, however, that low frequency measurements might be even more suitable to obtain information of the relaxation processes taking place in amorphous polymers at low temperatures.

EXPERIMENTAL

We have measured internal friction Q^{-1} and changes in the sound velocity down to 0.3 K using a vibrating reed device[7]. Our samples have been prepared from commercial acrylic glass by sawing ribbons of 0.15 mm x 4 mm x 30 mm from the bulk material. After covering both sides of the ribbons with a 10 nm thick gold layer, one end of each sample was clamped between two copper flats (see Fig. 1). The free end vibrates under the influence of the electrostaticly applied sinusoidal voltage. In order to measure the velocity of sound the frequency of the applied voltage has been matched to the frequency f_n of one of the flexural eigenmodes, which are given by

$$f_n = a_n \frac{d}{l^2} \sqrt{\frac{E}{\rho}} \qquad (1)$$

Here a_n is a constant, E is Young's modulus, ρ the mass density, d the thickness of the sample and l its length.

By monitoring the resonance frequency as a function of temperature one can measure the change in the sound velocity $v_E = (E/\rho)^{1/2}$ which is related to the velocity v_l of longitudinal sound waves through the relation

$$v_l^2 = v_E^2 \frac{1-\sigma}{(1+\sigma)(1-2\sigma)} \qquad (2)$$

where σ is Poisson's ratio, so that $v_l = 1{,}23\ v_E$. The amplitude of the vibrating reed is detected by a second electrode placed opposite to the driving electrode. With a constant driving voltage the variation of the vibrational amplitude reflects the variation of the internal friction Q^{-1}. Its absolute value has been obtained by two different techniques. For relatively high damping factors the magnitude of Q^{-1} has been determined from the half-width of the resonance curve when slowly sweeping the frequency of the driving voltage through mechanical resonance. For relatively low damping factors the free exponen-

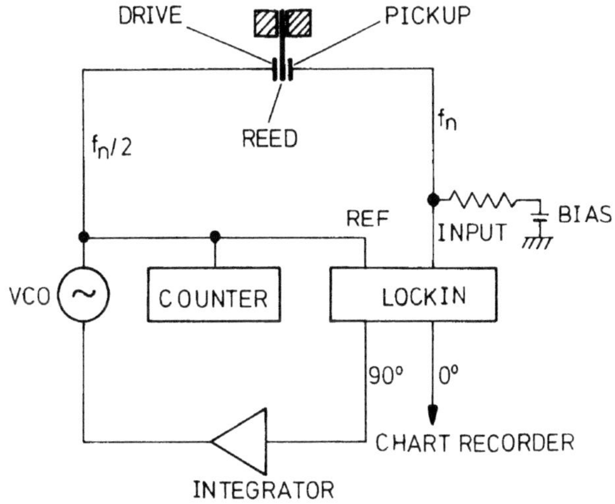

Fig. 1. Schematic of the experimental set up.

tial decay of the vibrational amplitude has been measured after switching off the driving force.

In order to achieve high precision the reed has been phase-locked to one of its natural frequencies f_n by tuning a voltage-controlled oscillator with the smoothed output of a lock-in amplifier (see Fig. 1). The frequency has been measured with an averaging counter with a resolution of 10^{-7}. The amplitude output of the lock-in amplifier yields relative changes in Q^{-1} to better than 1%.

RESULTS

The temperature dependence of the resonance frequency of our PMMA sample is shown in Fig. 2. From 0.3 K up to room temperature it decreases steadily by about 20%. Because thermal expansion is comparably small in this temperature range[8], the variation of the frequency reflects directly the temperature dependence of the sound velocity v_E.

In Fig. 3 the low temperature variation of the velocity is shown in more detail. Clearly in the temperature range 0.3 K to 1.5 K the velocity decreases logarithmically with temperature. Above 2 K the variation becomes more and more pronounced.

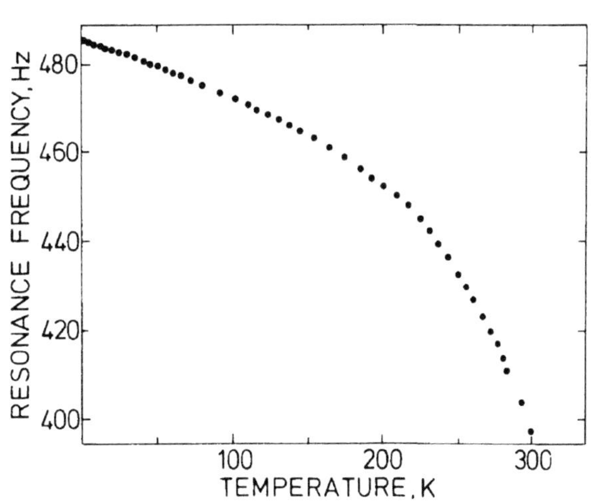

Fig. 2. Temperature dependence of the resonance frequency.

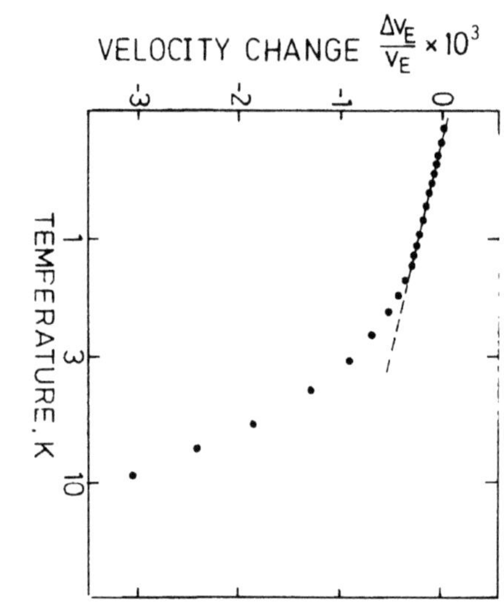

Fig. 3. Temperature dependence of the sound velocity v_E below 10 K. The dashed line indicates the logarithmic variation at the lowest temperatures.

In Fig. 4 the internal friction Q^{-1} is plotted as a function of temperature. Up to 6 K a distinct plateau is observed. Only a slight decline from $5 \cdot 10^{-4}$ at 0.3 K to $4.6 \cdot 10^{-4}$ at 6 K occurs, a feature which has also been found in inorganic glasses[6]. Around 30 K the damping passes through a minimum and rises steeply again. The high temperature behaviour is shown in Fig. 5 on a logarithmic scale. The main feature is an exponential increase of Q^{-1} with temperature up to room temperature with no indication of levelling off for the expected peak at higher temperatures. As indicated by the hatched area a broad loss peak is superimposed on the exponential rise in the temperature range from 40 K to 140 K.

THEORY

A closer inspection of our results reveals striking similarities between the low temperature elastic properties of inorganic glasses and those of PMMA. As mentioned earlier this is also true for thermal properties. Therefore we use the concept of the so-called "Tunneling

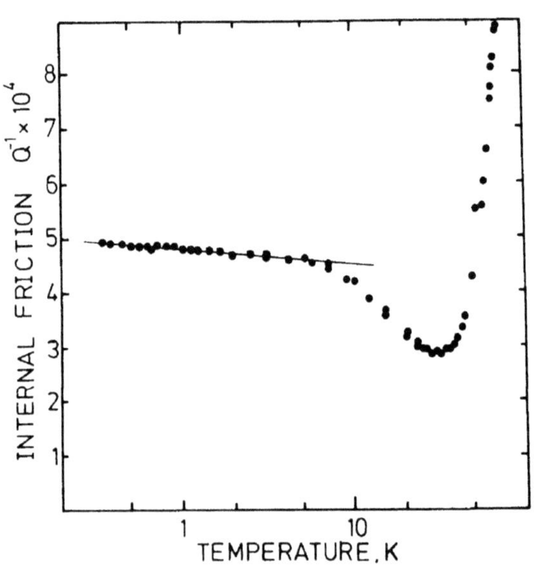

Fig. 4. Temperature dependence of the internal friction Q^{-1}; the straight line is a visual fit indicating the "plateau".

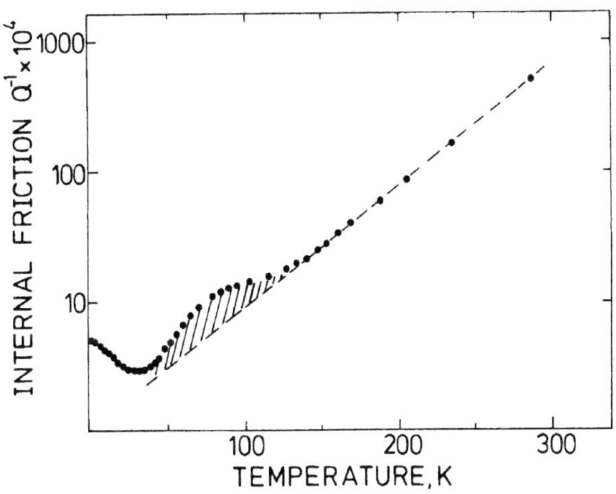

Fig. 5. Temperature dependence of the internal friction Q^{-1}. The dashed line shows the exponential increase, the hatched area the superimposed loss peak.

Model"[9, 10], which has been very sucessfully applied to the description of the cryogenic anomalies of inorganic amorphous solids.

According to this model the random amorphous structure enables certain small clusters of atoms to occupy two energetically nearly equivalent positions of equilibrium which correspond to the minima of the double-well potential shown in Fig. 6. Tunneling between the two positions may occur if the wavefunctions of the particle in either well overlap. The parameter describing the overlap is the so-called tunneling parameter λ, which can be written as

$$\lambda = \frac{d}{\hbar}\sqrt{2mV} \tag{3}$$

where d represents the separation of the two wells (in real space or in configurational space), m the effective mass and V the potential barrier as shown in Fig. 6. The tunneling motion of the particle leads to a splitting of the ground states given by

$$\Delta_0 = \hbar\Omega e^{-\lambda} \tag{4}$$

where $\hbar\Omega$ is of the order of the zero point energy of the particle. If furthermore the two minima are not exactly equivalent, then the energy

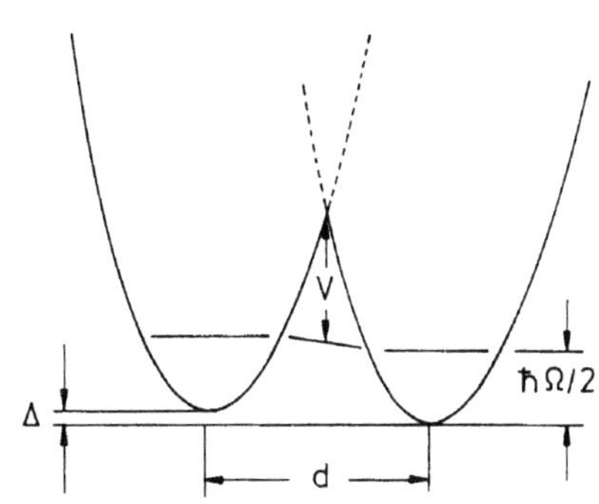

Fig. 6. The double-well potential of the Tunneling Model with barrier height V, asymmetry Δ and distance d between the two minima. $\hbar\,\Omega/2$ is the ground state energy of the tunneling particle.

splitting of the resulting two-level system is given by

$$E = \sqrt{\Delta^2 + \Delta_0{}^2} \tag{5}$$

Because of the random nature of the amorphous structure the parameters characterizing the two-level tunneling states will not exhibit well-defined values but rather a broad distribution. In the "Tunneling Model" the quantities Δ and λ are assumed to be independent and uniformly distributed, i.e. we may use the simple distribution

$$P(\Delta,\lambda)\ d\Delta d\lambda = \bar{P}\ d\Delta d\lambda \tag{6}$$

where $\bar{P}$ has a constant value.

Let us now have a short look at the dynamical behaviour of the tunneling systems. After an external perturbation of their population, they will reattain thermal equilibrium within a certain relaxation time τ. In dielectric glasses thermal equilibrium of the ensemble of tunneling states is established via the emission or absorption of thermal phonons. At low temperatures the one-phonon process is dominating the relaxation rate[11] and we may write

$$\tau^{-1} = A\left(\frac{\Delta_0}{E}\right)^2\left(\frac{E}{k}\right)^3 \coth\frac{E}{2kT} \tag{7}$$

where A is a characteristic constant. Its value depends on the material and is about $1.2\cdot10^8 K^{-3}s^{-1}$ in vitreous silica and $1.6\cdot10^9 K^{-3}s^{-1}$ in the case of PMMA. It is important to note that the relaxation rate depends on the ratio Δ_0/E, meaning that for a fixed energy value E the relaxation rate is a function of Δ_0. For completely symmetric systems the energy splitting is due to the tunneling motion only, i.e. $\Delta_0/E = 1$ (see Eq. 5). In this case the relaxation time exhibits its minimum value $\tau_m(E)$. Using Eqs. 5 and 7 we may rewrite the spectral distribution given by Eq. 6 with E and τ^{-1} as variables.

We obtain for $\tau>\tau_m(E)$

$$P(E,\tau^{-1}) = \frac{\bar{P}}{2}\ \frac{\tau}{(1-\tau_m(E)/\tau)^{1/2}} \tag{8}$$

This function is plotted for a fixed value of E in Fig. 7. Note that the function tends to infinity for the extreme values of τ.

Using this model, what do we expect for the elastic low temperature properties of PMMA? For our frequency and temperature range (40 Hz<f<1 kHz and T>0.3 K, respectively) we always fulfil the condi-

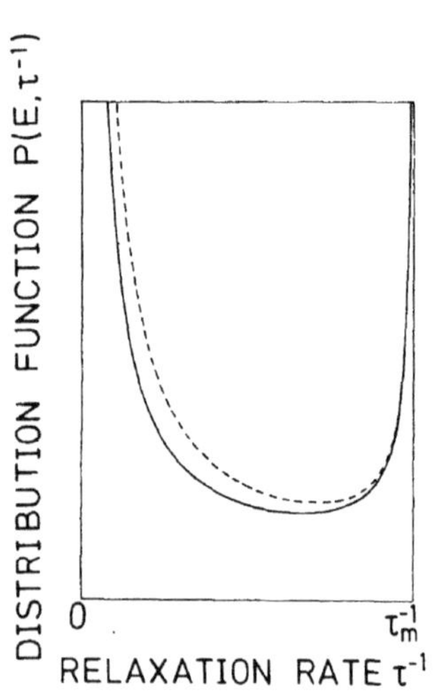

Fig. 7. Distribution function $P(E,\tau^{-1})$ in arbitrary units as given by Eq. 8 (full line); dashed line see text.

tion $\omega\tau_m \ll 1$. In this limiting case the relaxation process gives rise to a variation of the sound velocity given by[12]

$$\left(\frac{\Delta v}{v_0}\right)_{rel} = -\frac{3}{2} C \ln \frac{T}{T_0} \qquad (9)$$

where v_0 is the sound velocity at an arbitrary reference temperature T_0 and C is a constant characteristic for the material. However, an additional attenuation mechanism contributes to the experimentally observed temperature variation of the velocity. It is well known that the tunnelling states existing in amorphous solids can also absorb phonons resonantly. This process also leads to a logarithmic variation of the velocity. It has opposite sign and its magnitude is just 2/3 of the value given in Eq. 9. Thus the two processes add up to a variation of the sound velocity given by

$$\frac{\Delta v}{v_0} = -\frac{1}{2} C \ln \frac{T}{T_0} \qquad (10)$$

Although we have in principle two contributions to the internal friction, namely the resonant and the relaxation absorption, we see only the latter one in our experiment. Because of the frequency dependence of the resonant absorption this contribution can be neglected in our frequency range. For the relaxation process we expect[10]

$$Q^{-1} = \frac{\pi}{2} C \qquad (11)$$

This means that the internal friction should be independent of temperature and frequency and is therefore often called "plateau". The characteristic constant is given by

$$C = \frac{\bar{P}\gamma^2}{\rho v^2} \qquad (12)$$

where $\bar{P}$ is the constant spectral density (see Eq. 6) and γ is the coupling constant (or deformation potential) describing the interaction between two-level tunneling states and strain fields.

DISCUSSION

As pointed out previously the predictions of the tunneling model have been verified in inorganic glasses in the temperature range 0.02 K to 2 K. PMMA shows the expected logarithmic decrease of the velocity with temperature (Eq. 10) up to temperatures of 1.5 K. From the measured slope we deduce $C = 4.2 \cdot 10^{-4}$. This value is in good agreement with the value $C = 4 \cdot 10^{-4}$ obtained from measurements of the velocity of sound at 18 GHz by Brillouin scattering[5]. Above 1.5 K the velocity varies much more rapidly with temperature. A similar behaviour has also been found for inorganic glasses and has been explained as the onset of higher order phonon processes in the relaxation of the tunneling states[6]. So far no explicit theoretical model has been developed to explain the elastic properties in this temperature range.

Internal friction is almost temperature independent up to 6 K. From the measured magnitude of Q^{-1} we find $C = 3 \cdot 10^{-4}$. This value is substantially smaller than that deduced from the sound velocity measurements. A similar discrepancy as well as the slight decline of the "plateau" is also seen in other glasses[6] and might therefore be a more general feature.

Better agreement between the values of the coupling constants C deduced from experiment could be achieved by slightly modifying the original distribution function as given by Eq. 6. For example, it would be sufficient to increase the number of slowly relaxing, weakly coupled tunneling states with respect to these having a high Δ_0/E ratio. This artificial shift is indicated by a dashed line in Fig. 7.

Around 6 K the "plateau region" of the internal friction ends, indicating that the original Tunneling Model is not valid anymore. Since in the relaxation absorption those tunneling states contribute most to the damping which have an energy splitting comparable to 1.5 kT, one might argue that for E/k>10K the density of states does not remain constant anymore. Again a slight modification, but now in the energy dependence of $P(E,\tau^{-1})$ would lead to a reasonable agreement between theory and experiment. Here we do not pursue this idea further, since changing the original density of states without knowing the underlying physical origin of this modification does only add one more free parameter without leading to a deeper understanding of the phenomenon.

The superimposed loss maximum at about 70 K (Fig. 5) is probably due to the presence of small amounts of dibutylphtalate (DBP) - a plasticizer - or higher polymethacrylate esters[1]. Studies on purified samples are in progress.

The exponential increase of the internal friction with increasing temperature is quite amazing because at these temperatures one would expect a thermally activated process, i.e. an Arrhenius behaviour. It predicts a linear relation, between log Q^{-1} and T^{-1} with a slope proportional to the barrier height. Using this picture our data could be fitted, if the activation energy were proportional to T^2, meaning that the underlying process is not a simple hopping over barriers.

CONCLUSION

We have shown that low frequency measurements are suitable for the observation of the logarithmic decrease of sound velocity with increasing temperature and the frequency and temperature independent plateau of the internal friction predicted by the tunneling model. Below 6 K PMMA exhibits the same mechanical behaviour as inorganic glasses, chemistry or short range order do not play an important role. As in the case of inorganic glasses the tunneling particles cannot be identified yet. However, PMMA and other amorphous polymers are promising tools, since organic chemistry supplies a great variety of possible candidates for the tunneling entities.

REFERENCES

1. J. Heijboer and M. Pineri, Dynamic mechanical properties of polymethacrylates at low temperatures, in: "Nonmetallic Materials and Composites at Low Temperatures", vol. 2, G. Hartwig and D. Evans, eds., Plenum Press, New York (1982), pp. 89-116.
2. J. Heijboer, Molecular origin of relaxations in polymers, in: "The Glass Transition and the Nature of the Glassy State", M. Goldstein and R. Simha, eds., Ann. N.Y. Acad. Sci., vol. 279, 104 (1976).
3. R. B. Stephens, Intrinsic low-temperature thermal properties of glasses, Phys. Rev. B 13, 852 (1976).
4. G. Federle and S. Hunklinger, Ultrasonic studies of some polymers at low temperatures, J. Physique 43, C9, 505 (1982).
5. M. Schmidt, R. Vacher, J. Pelous and S. Hunklinger, Brillouin scattering from amorphous polymers at low temperatures, J. Physique 43, C9, 501 (1982).
6. A. K. Raychaudhuri and S. Hunklinger, Low frequency elastic properties of glasses at low temperatures - implications on the

tunneling model, to be published in Z. Phys. B - Condensed Matter 57 (1984).
7. B. S. Berry, W. C. Pritchet, Vibrating reed internal friction apparatus for films and foils, IBM J. Res. Dev. 19, 334 (1975).
8. K. G. Lyon, G. L. Salinger and C. A. Swenson, Tunneling models and the experimental thermal expansivities of fused silica and polymethylmethacrylate (PMMA) below 4 K, Phys. Rev. B 19, 4231 (1979).
9. W. A. Phillips, Tunneling states in amorphous solids, J. Low Temp. Phys. 7, 351 (1972).
10. J. Jäckle, On the ultrasonic attenuation in glasses at low temperatures, Z. Physik 257, 212 (1972).
11. J. Jäckle, L. Piché, W. Arnold and S. Hunklinger, Elastic effects of structural relaxation in glasses at low temperatures, J. Non-Cryst. Sol. 20, 365 (1976).
12. P. Doussineau, C. Frénois, R. G. Leisure, A. Levelut and J.-Y. Prieur, Amorphous-like acoustical properties of Na doped β-Al_2O_3, J. Physique 41, 1193 (1980).

VERY LOW FREQUENCY LOSS MEASUREMENTS IN GLASSES

Manfred von Schickfus

Institut für Angewandte Physik II
Universität Heidelberg
6900 Heidelberg FRG

Herbert Tietje and Eberhard Gmelin

Max-Planck-Institut für Festkörperforschung
7000 Stuttgart FRG

INTRODUCTION

At low temperatures the thermal, elastic and dielectric properties of glasses are known to be determined by low-energy excitations which seem to be intrinsic to the glassy state /1/. Recently measurements of the time dependence of the specific heat /2,3,4/ have shown that some of these excitations relax very slowly into thermal equilibrium. Such a behaviour, especially a logarithmic time dependence of the specific heat, is indeed predicted by the tunneling model /5,6/ which until now was most successful in explaining these low-temperature anomalies. A quantitative analysis of these time-dependent thermal measurements, however, is very difficult and does not allow to unambiguously confirm the predictions of the tunneling model.

EXPERIMENTAL

We have measured the elastic absorption of glasses at low temperatures and at very low frequencies. Between 0.5 K and 6 K and at frequencies from 10^{-2} to 25 Hz this absorption is determined by the relaxation of the tunneling states. In our experiment the loss was measured through the elastocaloric effect /7/. This method has several advantages: Compared with torsional pendulum measurements it is relatively simple and, due to the non-resonant character of this technique, broad frequency ranges can be swept. Since the sample is subject to longitudinal and uniform stress, the analysis of the data is straightforward.

Fig. 1 shows our apparatus. We used a ^{3}He evaporation cryostat which was immersed into a liquid helium bath. The samples were either a cylindrical rod 4ϕ x 120 mm of vitreous silica or a ribbon of metallic glass with dimensions 140 x 1,5 x 0.05 mm. At the lower end the samples were clamped and thermally anchored to the ^{3}He chamber. In the case of vitreous silica this was achieved with Invar mountings glued to the sample. The samples were surrounded by a radiation shield from which a stranded copper wire led to the upper mounting of the sample. In this

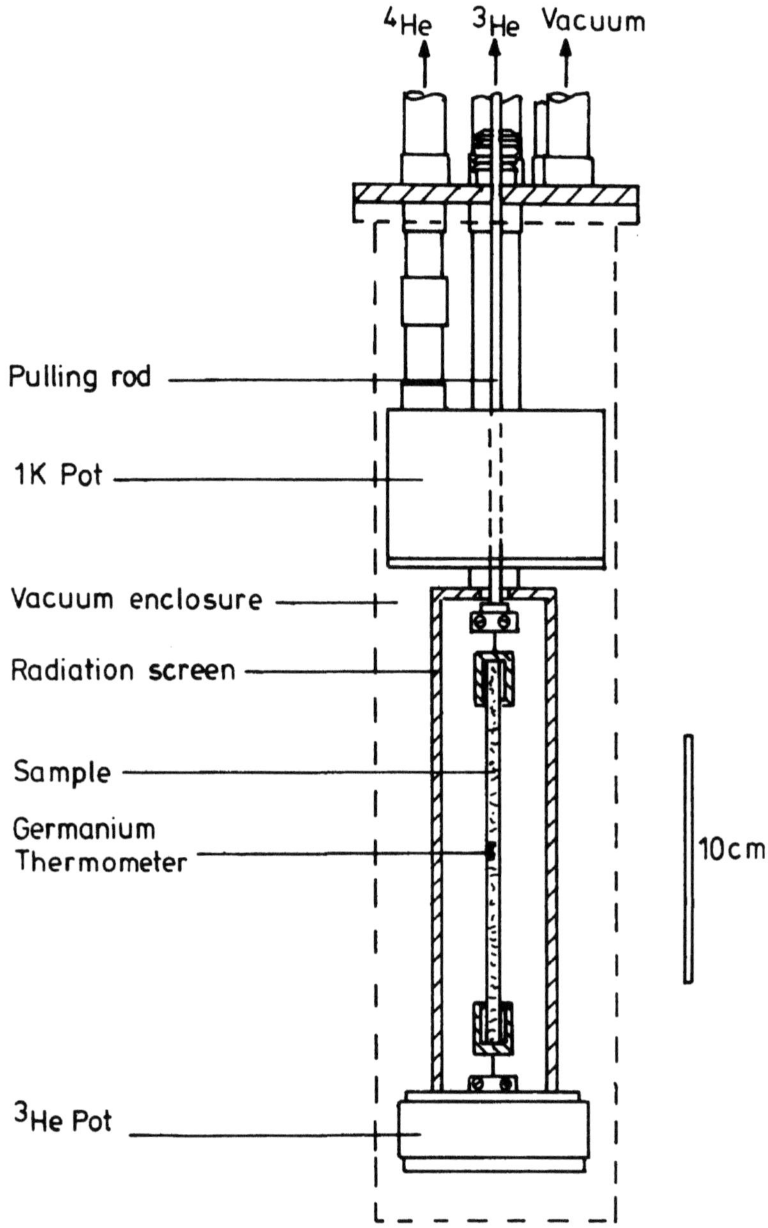

Fig. 1 Apparatus for low-temperature elastocaloric measurements. Only the lower end is shown, which is immersed into liquid helium. Strain is transmitted from outside the cryostat with a pulling rod.

way both ends of the sample were kept at the temperature of the ^{3}He chamber. Force was applied to the sample from outside the cryostat through a pulling rod, a thin-walled stainless steel tube. Sealing of this rod against the helium bath and against atmosphere was achieved with bellows in order to avoid friction. The pulling force was generated with an electrodynamic transducer (Bruel & Kjaer Mod. 4809), which could produce forces of ± 60N. This system was excited with a synthesized function generator (Philips PM 5190) and a power amplifier (Kepco BOP 20-10M). Force was measured with a strain gauge near the transducer, corrections for the influence of the bellows were determined at room temperature.

The experimental temperature was stabilized electronically to less than 1 mK with an additional thermometer and a heater at the ^{3}He chamber. Temperature variation due to the applied stress was measured in the middle of the sample with an exposed-element germanium resistor with a very low thermal mass. The resistance was measured with a simple Wheatstone bridge and a Lock-In Amplifier as a detector. The deviation of the bridge due to temperature variations was digitally recorded with a desktop computer. Absolute values for the amount of heat generated in the sample were obtained by resistive heating of the sample with a known power. For this purpose a thin Constantan wire was glued to the vitreous silica sample, whereas the metallic glass samples were heated by passing a current through them.

ELASTOCALORIC EFFECTS

If a sample with an internal friction Q^{-1} is subject to a periodic stress $\sigma = \sigma_o \cos \omega t$, there are two mechanisms which cause a temperature variation: The first is associated with the elastic loss and generates heat at a rate $\dot{q}_{loss}$ given by

$$\dot{q}_{loss} = (1/2Y)\omega\sigma_o^2 \, Q^{-1} \, (1 + \cos 2\omega t) \qquad (1)$$

where Y is Young's modulus. Therefore one expects a temperature rise of the sample which is modulated at twice the excitation frequency and which - at constant absorption - should increase linearly with frequency. The second mechanism is due to the anharmonicity of the elastic forces and can therefore be described with the Grüneisenparameter:

$$\dot{q}_{ah} = 1/3 \, \Gamma \, c_v \chi T \omega \sigma_o \, \sin\omega t \qquad (2)$$

where χ is the compressibility and c_v the specific heat of the sample. This contribution oscillates at the excitation frequency and causes a periodic cooling and heating of the sample. This thermoelastic effect can be regarded as the inverse of a normal thermal expansion experiment: here the length of the sample is varied and the corresponding change of temperature is measured. Since the contributions of equs. (1) and (2) appear simultaneously, the ω- and 2ω-contribution were separated by digital filtering techniques in our experiment.

RESULTS AND DISCUSSION

Fig. 2 shows the temperature response of a sample of vitreous silica together with the applied stress at three temperatures. It can easily be seen that at the highest temperature (6K) the anharmonicity (equ. 2) of the sample dominates the thermal response of the sample, whereas at low temperatures the main contribution comes from internal friction (equ. 1). Similar results have been obtained with the metal-

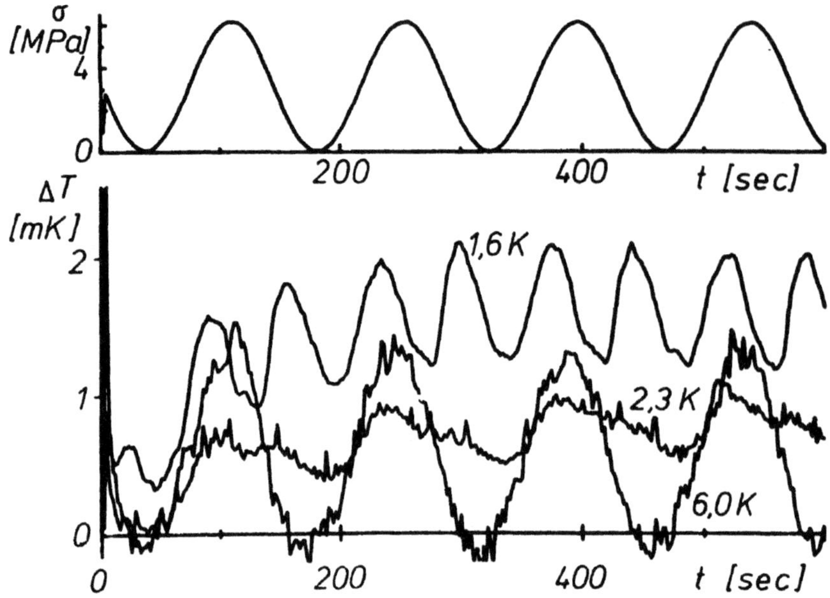

Fig. 2 Thermal response of vitreous silica at three different temperatures. The applied strain is shown on top. At 6K the anharmonicity of the sample dominates with a response at the excitation frequency, whereas at the lowest temperature absorption leads to periodic heating at twice the excitation frequency.

lic glasses PdSiCu and PdZr.

The tunneling model predicts a distribution of relaxation times of the tunneling states /8/

$$p(E,\tau^{-1}) = \bar{P}/2\tau(1 - \tau_m/\tau)^{-1/2} \tag{3}$$

where τ_m is a minimal relaxation time at a given temperature and energy E of the tunneling state. For $\omega \ll \tau_m^{-1}$ the contribution of this relaxation to the elastic loss can be calculated as /8/

$$Q^{-1} = n\bar{P}\,\gamma^2/(2\rho v^2) \tag{4}$$

where $\bar{P}$ is the density of the tunneling states and γ a deformation potential describing their coupling to phonons. ρ is the mass density of the sample and v the longitudinal sound velocity.

According to equs. (4) and (1), $\dot{q}_{abs}$ is expected to be temperature independent and to increase linearly with temperature. Figs. 3 and 4 show the corresponding results for vitreous silica: The heat produced by the elastic loss does increase linearly with frequency over more than three decades between 0.05 and 10 Hz and it remains almost temperature independent between 0.5 and 6 K. Qualitatively similar results are found for the amorphous metals PdSiCu and PdZr. The product of the density of

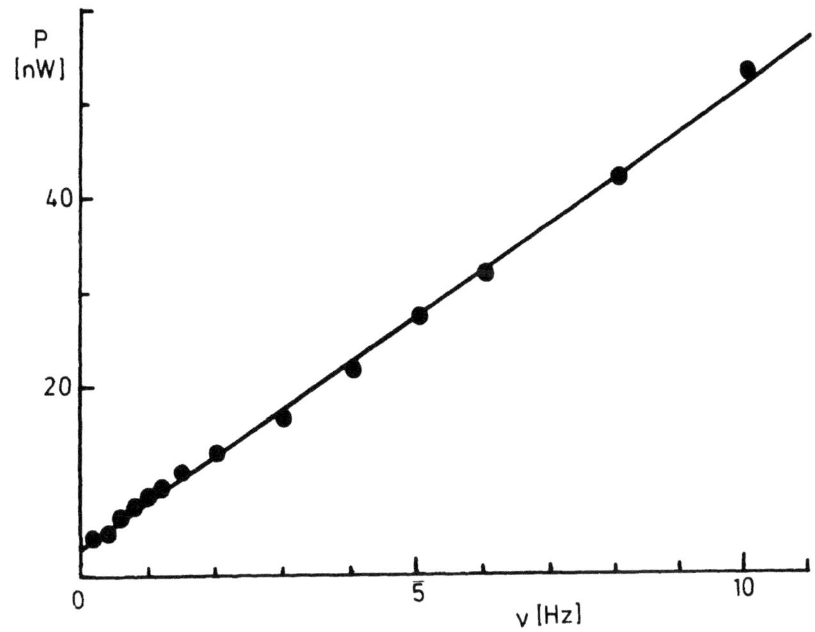

Fig. 3 Frequency dependence of the heat $\dot{q}_{abs}$ produced by the absorption process in vitreous silica at a temperature of 3.6 K.

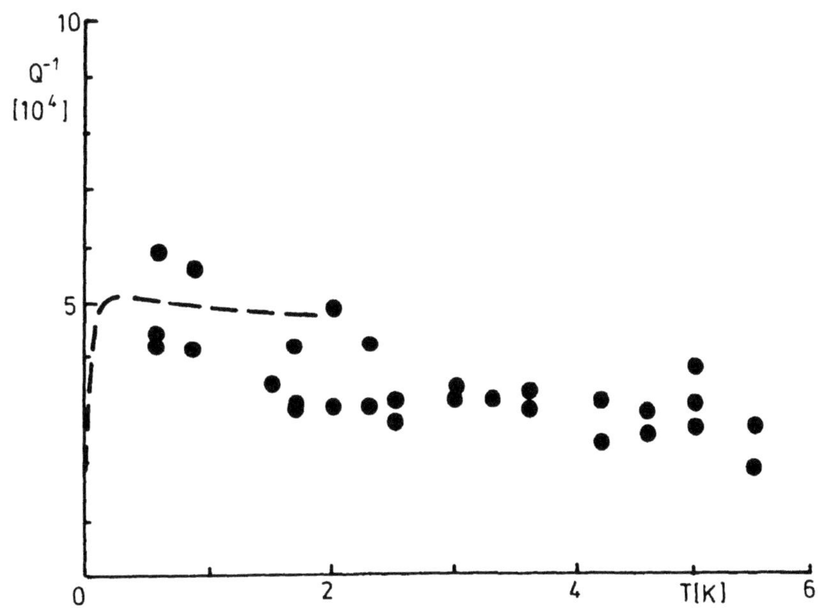

Fig. 4 Temperature dependence of the absorption Q^{-1} in vitreous silica. Absorption decreases only slightly with temperature. The dashed line indicates results from vibrating reed measurements at 1 kHz /12/.

states $\bar{P}$ and the square of the deformation potential γ (equ. 4) can be calculated from our absorption values and is shown in table 1 for all three materials investigated. This quantity can also be derived from sound velocity measurements around 100 MHz, these values are shown for comparison.

Since the main contribution to the absorption comes from tunneling states for which $\tau^{-1} \simeq \omega$ and $E \simeq kT$, our results at very low frequencies show that the distribution of τ^{-1} follows the prediction of equ. 3. The range of τ^{-1} must extend over more than 10 decades at a given energy since $\tau_m^{-1} \simeq 10^9 s^{-1}$ for our experimental situation with $T \simeq 2K$ and E kT. We find therefore are an excellent confirmation of the assumptions of the tunneling model.

TABLE 1

	$\bar{P}\gamma^2$	
	this experiment	sound velocity
Vitreous Silica Suprasil I	$1.7 \cdot 10^8$ erg/cm^3	$2.1 \cdot 10^8$ erg/cm^3 /9/
$Pd_{0,775}Si_{0,165}Cu_{0,06}$	$7.6 \cdot 10^7$ erg/cm^3	$6,4 \cdot 10^7$ erg/cm^3 /10/
$Pd_{0,3}Zr_{0,7}$	$2.9 \cdot 10^7$ erg/cm^3	$7 \cdot 10^7$ erg/cm^3 (700MHz) /11/

CONCLUSION

With a calorimetric method we have measured the elastic loss of dielectric and metallic glasses at low temperatures and at very low frequencies. Our results confirm the prediction of a distribution of relaxation times made by the tunneling model and show that this distribution must extend over at least 10 orders of magnitude.

Finally we would like to mention that our method is not limited to the study of amorphous materials. We think that it is a very useful and simple technique for the study of the low-temperature loss in a large variety of materials.

REFERENCES

1. For a recent review see, for example "Amorphous Solids-Low Temperature Properties", W.A. Phillips ed., Springer-Verlag, Berlin (1981)
2. M. Meissner, K. Spitzmann, Experimental Evidence on Time-dependent Specific Heat in Vitreous Silica, Phys. Rev. Lett. 46, 265-268 (1981)
3. M.T. Loponen, R.C. Dynes, V. Narayanamurti, J.P. Garno, Measurement of the Time-dependent Specific Heat of Amorphous Materials, Phys. Rev. B25, 1161-1173 (1982)
4. J. Zimmermann, G. Weber, Thermal Relaxation of Low-Energy Excitations in Vitreous Silica, Phys. Rev. Lett. 46, 661-664 (1981)
5. W.A. Phillips, Tunneling States in Amorphous Solids, J. Low-Temp. Physics 7, 351-360 (1972)
6. P.W. Anderson, B.I. Halperin, C.M. Varma, Anomalous Low-Temperature Thermal Properties of Glasses and Spin Glasses and Spin Glasses, Phil.Mag. 25, 1-9 (1972)

7. M.v. Schickfus, S. Hunklinger, K. Dransfeld, Elastic Nonlinearity of Vitreous Silica at Low Temperatures, in:"Non metallic Materials and Composites at Low Temperatures", vol. 2, Planum Press, New York (1982), pp. 37-48

8. J. Jäckle, On the Ultrasonic Attenuation in Glasses at Low Temperatures, Z. Physik 257, 212-223 (1972)

9. S. Hunklinger, W. Arnold, Ultrasonic Properties of Glasses at Low Temperatures, in "Physical Acoustics", vol. 12, W.P. Mason, R.N. Thurston, eds., Academic Press, New York (1976), pp. 155-215

10. B. Golding, J.E. Graebner, A.B. Kane, J.L. Black, Relaxation of Tunneling Systems by Conduction Electrons in a Metallic Glass , Phys. Rev. Lett. 41, 1487-1491 (1978)

11. G. Weiß, S. Hunklinger, H. v. Löhneysen, Influence of Superconductivity on the Sound Propagation in Disordered Metals. Physica 109B&110B, 1946-1954 (1982)

12. A.K. Raychaudhuri, S. Hunklinger, Low Frequency Elastic Properties of Glasses at Low Temperature, J.de Physique 43, C9-485-492

MECHANICAL AND DIELECTRIC LOSSES OF POLYMERS IN THE TEMPERATURE RANGE OF 5K TO 293K

G. Hartwig and G. Schwarz

Nuclear Research Center Karlsruhe, IRB/ZSM

Karlsruhe, Federal Republic of Germany

INTRODUCTION

Technical applications and scientific research are main reasons for studying losses of polymers at low temperatures.

The technological aspect is: Heat generation or energy dissipation caused by alternating mechanical or dielectric loading might become a non-negligible parameter in the costs of cooling with liquid helium.

The scientific aspect is: Damping and the relaxation behavior of polymers are useful in analysing molecular motions. At low temperatures most degrees of freedom are frozen and can be activated separately. Current theories explain damping as changes of place of molecular segments or side groups induced by external mechanical or electric excitation.

In measurements with oscillating loads a maximum of damping occurs when the external oscillating time $1/\omega$ (inverse frequency) is equal to the respective relaxation time τ[1,2].

$$\tau(T)\cdot\omega = 1 \qquad (1)$$

Usuallly, damping is expressed by the loss tangent which describes the ratio of the dissipative energy and the stored energy, both related to one cycle. The objective of this investigation is the correlation of mechanical and dielectric properties.

They were determined:

a) in the mechanical case with a torsion pendulum in a cryogenic version at a roughly constant low frequency (8 Hz - 15 Hz).

b) in the electric case with a Schering bridge at 50 Hz.

All experiments were performed at temperatures from 5 K to 293 K. The real and imaginary parts of the complex shear modulus and the complex dielectric permittivity were determined.

The results obtained showed an asymptotic behavior at very low temperatures and a surprisingly good proportionality of the mechanical to dielectric loss tangents at temperatures below 200 K.

POLYMERIC MATERIALS

Six polymers were selected for measurement with regard to their polarity, crystallinity and type of molecular motion. The materials and some of their properties are listed in Table I. The chemical symbols and some dipole moments of polar groups (arrows) are given in addition[3].

High density polyethylene: It is a nearly unpolar polymer. A single C-H bond has a dipole momentum of about 0.4 D.U. (1 Debye unit is equal to 3 x 10^{-30} C·m). On account of the well-ordered configuration of these groups cancellation of surrounding dipoles results in an effective mean polarity of the order of less than 0.03 D.U. per CH_2-mer. The polymer has a crystallinity of about 65 vol %.

Polystyrene: This polymer is weakly polar with a mean polarity of 0.4 D.U.. It is a linear amorphous polymer. The side groups are phenyl rings with an atactic arrangement.

Polycarbonate: It has a linear structure with a weak crystallinity. The dipole moments are located at the intrachain phenyl rings, the methyl groups and the oxygen atoms. This totals up to a mean polarity of about 1 D.U..

Epoxy resins: The epoxy materials are amorphous with a mean polarity of about 1.2 D.U. They have only small molecular side groups like OH- or methyl groups. Both epoxy resins selected differ by their cross-link density. My 740/D230 has the higher density.

Polyvinyl chloride: PVC is a strongly polar polymer with a mean polarity of about 1.5 D.U. due to its C-Cl mers. The commercial systems have an atactic arrangement of the chloride atom; thus, little cancellation takes place there.

Table I. Description of Polymers Selected

Polymers	Polarity	Morphology	Side Groups	Chemical Symbols
High density polyethylene HDPE	unpolar $\bar{\mu} \approx 0.03$ D.U. cancellation	partly cryst. ≈ 65 vol. %	None	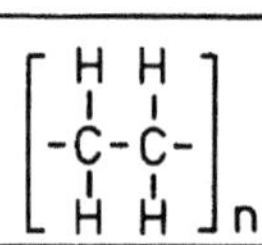
Polystyrene PS	weakly polar $\bar{\mu} \approx 0.4$ D.U.	amorphous	phenylring	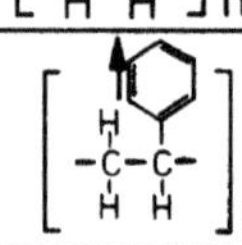
Poly-carbonate PC	medium polar $\bar{\mu} \approx 1$ D.U.	weakly crystalline	oxygen and methylgroups	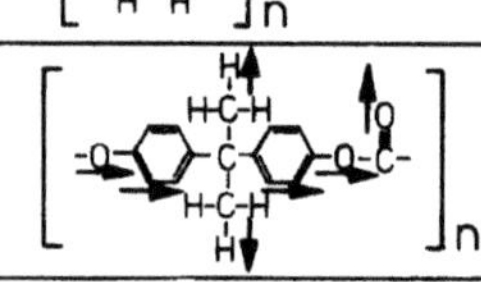
Epoxy resins Cy221/Hy979 My740/D230	polar $\bar{\mu} \approx 1.2$ D.U.	amorphous crosslinked	OH oxygen - methylgroups	
Polyvinyl chloride PVC	strongly polar $\bar{\mu} \approx 1.5$ D.U.	amorphous	Cl- atoms	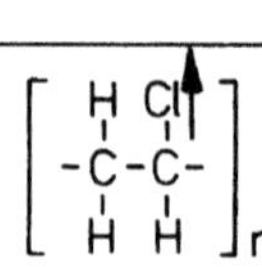

CRYOGENIC APPARATUS AND MEASUREMENTS

A torsion pendulum in a cryogenic version is a tool for determining mechanical losses at low temperatures. The loss factors are calculated with the measured logarithmic decrement Λ ($\tan\delta = \Lambda/\pi$ for $\tan\delta \ll 1$). The torsional frequency, the geometric parameters, and the momentum of inertia are the parameters allowing to determine the shear modulus.

The dielectric properties were measured with a Schering bridge by detecting the capacity and the phase shift of the specimen, which is designed as a guard ring condensor. This guarantees a homogeneous electric field and minimizes errors. The boundaries of the specimen are shaped like a Rogowski profile in order to avoid corona effects and creep currents. Both surfaces of the specimen are painted with a conducting screen as electrodes.

During the cooling-down period both electrodes were grounded to inhibit freezing of the polarisation in a favoured direction like in an electret. High voltage is applied in data taking only. The measuring device permits to achieve a sensitivity of about 10^{-8} for $\tan\delta$.

POLYMERS UNDER ALTERNATING LOAD. THE COMPLEX MODULI

When alternating load is acting on a solid, the response generally lags behind by a phase angle δ. This phase lag results from the time necessary for molecular rearrangements. The loss properties are usually expressed by complex moduli. They are:

Complex Young's modulus: $E^* = E' + i\,E''$ (2)

Complex shear modulus: $G^* = G' + i\,G''$ (3)

The loss factor or loss tangent is defined by the ratio of the dissipated energy and the stored energy in one cycle.
It holds for the mechanical loss tangent:

$$\tan\delta_{mech} = \frac{G''}{G'} \simeq \frac{E''}{E'} \;^{**} \tag{4}$$

The complex dielectric permittivity is defined by:

$$\varepsilon^* = \varepsilon' - i\,\varepsilon'' \tag{5}$$

and the dielectric loss tangent is:[1]

$$\tan\delta_{el} = \frac{\varepsilon''}{\varepsilon'} \tag{6}$$

Analogous mechanical and dielectric properties are:

$$E^* \text{ or } G^* \text{ and } \varepsilon^{*-1}$$

**) Both moduli are correlated by the complex Poisson's ratio: $\mu^* = \mu' - i\mu''$ (7)
It holds:

$$\frac{E''}{E'} = \frac{G''}{G'} - \frac{\mu''}{1+\mu'} \tag{8}$$

In solids the last term in eq. (8) is negligible.

E' or G' and $(1/\varepsilon)'$

E'' or G'' and $(1/\varepsilon)''$

But it can be shown, that $\tan\delta_{mech} = \frac{G''}{G'}$ is analogous to $\tan\delta_{el} = \frac{\varepsilon''}{\varepsilon'}$ [1]

POTENTIAL BARRIERS AND MOLECULAR MOTIONS

In polymeric solids potential barriers have to be jumped over when changes of place occur which arise from the following binding potentials:

- COVALENT BINDING POTENTIAL (from quantum mechanical exchange forces). They determine the distance and the configuration of atoms within a polymeric molecule. Changes of configuration may take place by rotations[4]. The rotational potential as a function of the azimuthel angle is plotted in Fig. 1 for polyethylene. An absolute minimum of the energy only exists in the trans-position. The real rotational potential barriers in polymeric solids are rather high because of steric hindrance. Usually they can be overcome appreciably at rather high temperatures where the relaxation time is shorter. In polyethylene or other linear molecules like PVC, however, a collective motion of 4 or 5 mers around a collinear axis (crankshaft motion) takes place which reduces the effective rotational potential barrier.

- SECODARY BINDING POTENTIAL (e.g. Van der Waals potential)
They determine the mean distance of the molecules in a polymer. The main contributions arise from:[5]
a) the dispersion potential (quantum mechanical dipole exchange forces[6]). It forms the main quota of the secondary binding forces in unpolar polymers;
b) dipole-dipole potentials. They are additional binding forces of polar polymers;
c) dipole induced potential.
An example of change of place in a Van der Waals potential is given by the wagging of neighbouring phenyl rings of PS[7]. The small potential barrier as shown in Fig. 2 allows to fulfill the conditions $\omega\cdot\tau = 1$ at low frequencies and at low temperatures (40 K, 10 Hz). (Extending these considerations to molecular chains, three-dimensional dipole oscillators have to be taken into consideration[8]). Mechanical forces are transmitted by all of the interactions a), b) and c) leading to changes of the ORIENTATION of molecular groups. The electric load is transmitted by interactions with dipoles. This results in an ORIENTATIONAL POLARISATION which is small for unpolar polymers and large for polar polymers. But in addition an

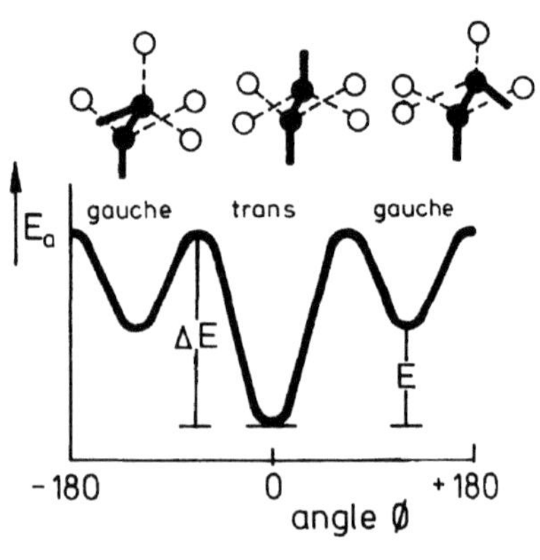

Fig. 1. Rotational potential of polyethylene (schematically)

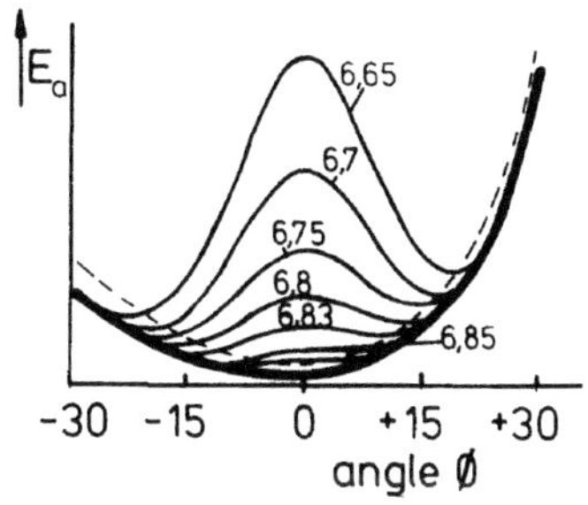

Fig. 2. Potential barrier at different relative distances of phenyl groups

electric field acts on the electron orbitals leading to an electron DEFORMATION POLARISATION. Its relaxation time is much too short for causing damping effects (imaginary part) in the low frequency experiment. Thus, only a contribution to the dielectric permittivity (real part) occurs, which is temperature independent at low frequencies[9].
So, the analogous mechanical and dielectric properties may be split as shown in Table II.

RESULTS AND DISCUSSION

The mechanical and dielectric loss tangents, shear modulus G' and inverse permittivity $(1/\varepsilon)'$ are plotted versus temperature in Figs. 3 to 8. The following results have been obtained:

1. The mechanical loss tangents of amorphous polymers approch nearly equal values at very low temperatures:

$$\text{amorphous polymers: } \tan\delta_{mech}\ (5\ \text{K}) \simeq 10^{-3}$$

Partially crystalline polymers have lower values depending on their crystallinity:

$$\text{part. cryst. polymers: } \tan\delta_{mech}\ (5\ \text{K}) \simeq 4 - 8 \cdot 10^{-4}$$

Naturally, values of the loss tangents will be higher if there is a relaxation at low temperatures:

Table II. ORIENTATIONAL (Or.) and DEFORMATION (Def.) contributions of Polarisation

	MECHANICAL				DIELECTRIC			
Polarisation	Def.		Or.		Def.		Or.	
Modulus	G'	G''	G'	G''	ε'	ε''	ε'	ε''
Unpolar	0	0	large	large	large	0	small	small
Polar	0	0	large	large	large	0	large	large
Loss tangents	$\tan\delta_{mech} = \frac{G''_{Or}(T)}{G'_{Or}(T)}$ (9)				$\tan\delta_{el.} = \frac{\varepsilon''_{Or}(T)}{\varepsilon'_{Or}(T) + \varepsilon'_{Def}(T)}$ (10)			

for example PS: $\tan \delta_{mech}$ (5 K) $\simeq 2 \cdot 10^{-3}$

2. Mechanical and dielectric damping peaks occur at the same temperature frequency points. Therefore, it can be concluded that the same molecular groups are activated both by electric and mechanical excitations and that in either case they exhibit the same relaxation behavior.

3. The dielectric and mechanical loss tangents differ by a factor C_1 which is independent of temperature and ranges from 10^{-2} (PE) to about 1 (PVC) for unpolar and polar polymers respectively, at temperatures

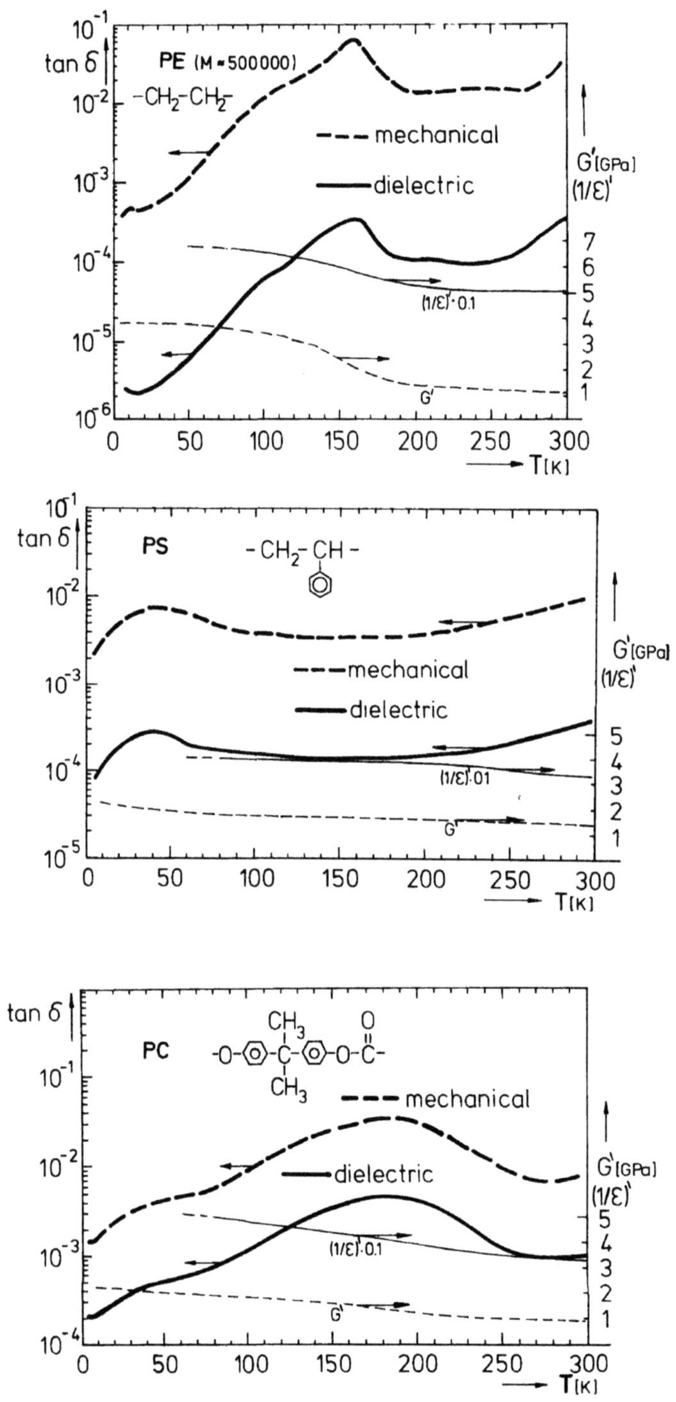

Figs. 3 to 5: Mechanical and dielectric loss tangents, G' and $(1/\varepsilon)'$ versus temperatures.

below 200 K.

$$\tan \delta_{el} = C_1 \tan \delta_{mech} \tag{11}$$

The great variation of C_1 belonging to the polymers considered is due to the different Or-contributions relative to the Def-contributions (see eq. (9), (10) in Table II. For PE this is not explained.)
The ratio of the real parts of the corresponding losses also leads to a constant, temperature independent factor C_2 except in the case of PE which will be considered separately.

$$\frac{(1/\varepsilon)'}{G'} = C_2 \tag{12}$$

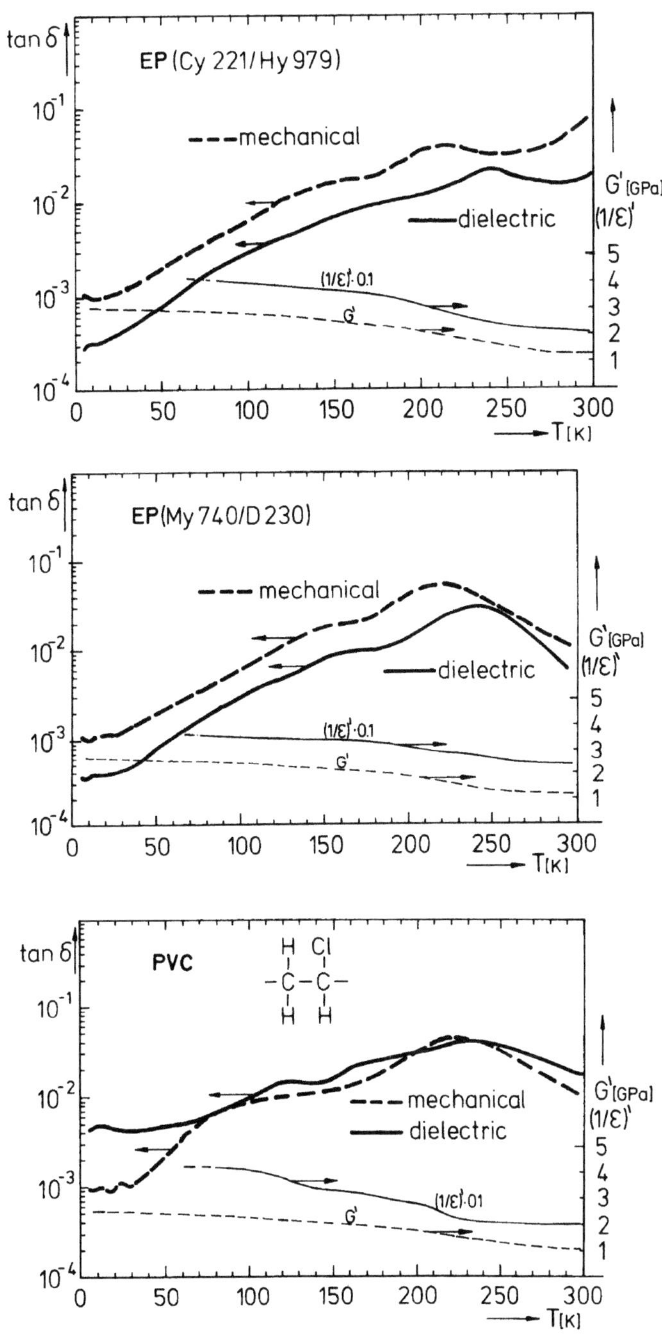

Figs. 6 to 8: Mechanical and dielectric loss tangents, G' and (1/ε)' versus temperatures.

4. Every damping peak causes a step of the real parts of the moduli, e.g. G' and $(1/\varepsilon)'$. The mechanical relaxation strength $\Delta G'$ in most cases is higher than the change of the dielectric permittivity $\Delta(1/\varepsilon)'$, especially for unpolar polymers. The change of G' and $(1/\varepsilon)'$ at a peak is due to random sequential mobilities in molecular chains. For PE (unpolar) nearly no change of $(1/\varepsilon)'$ as a function of temperature is expected to occur in the low-frequency range. But in the experiment a relatively great step was found at the secondary peak as seen from Fig. 9.

The low factor C_1 of about $2 \cdot 10^{-2}$ for PE cannot be explained especially since the relative large step $\Delta(1/\varepsilon)'$ in the permittivity indicates a relatively large Or.-contribution compared to the background arising from Def. polarisation. For an illustration of the problems the ratios of the mechanical and dielectric loss tangents (factor C_1) and those of the real parts (factor C_2) are plotted versus temperature in Fig. 9.

CONCLUSIONS

- Mechanical losses of amorphous polymers approach approximately the same value at 5 K.
  This is independent of the chemical structure but it depends on the morphology. A crystalline content shifts this value to lower levels; low temperature relaxation effects increases the value.

- Mechanical and dielectric loss peaks occur at the same temperature fre-

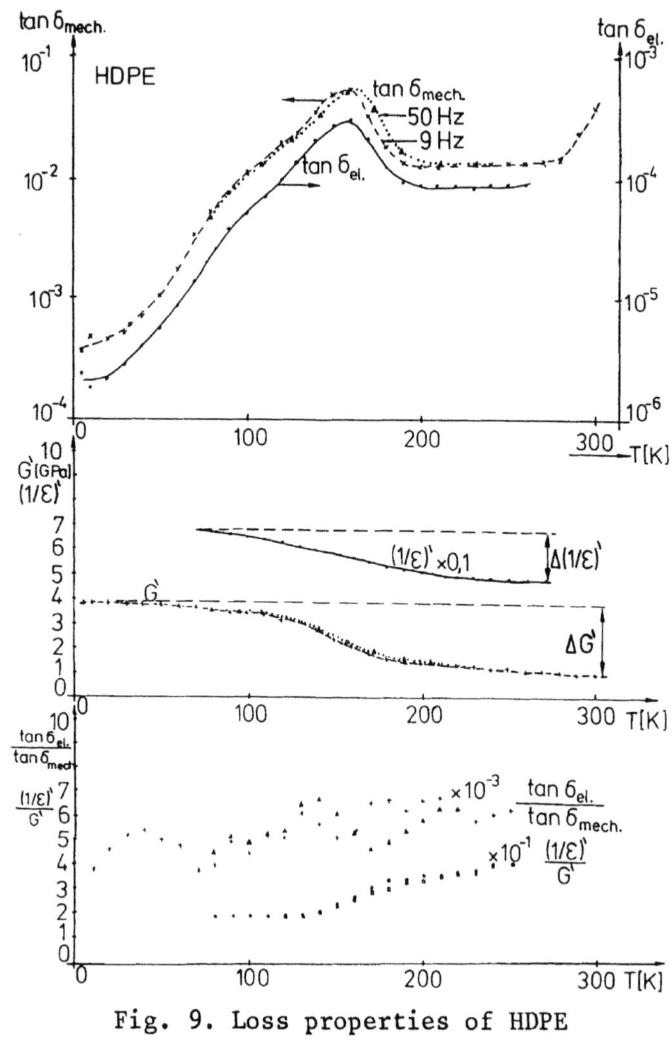

Fig. 9. Loss properties of HDPE

quency points which indicates a similarity in activation and relaxation behavior.

- A constant time-independent factor C_1, which is specific of each polymer and varies from 10^{-2} to 1, correlates mechanical and dielectric losses. A constant temperature independent factor C_2 correlates the real parts except for PE.
- Within the proportionality of the loss tangents the quotient of the mechanical and dielectric real parts is independent of temperature expect that of PE which needs further explanations.

APPENDIX

Dielectric and mechanical losses were determined at frequencies of the same magnitude. The torsional frequency for mechanical loss measurement was about 10 Hz whereas dielectric losses were determinded at 50 Hz. The temperature frequency relation of the Arrhenius equation was applied to shift the temperature of mechanical measurement to values corresponding to 50 Hz. The known potential barrier at a peak was used also in its vicinity. Below 200 K this shift causes a small error as can be seen in Fig. 9 for the damping peak of PE.

The transformed Arrhenius equation is[2]:

$$T_2 = \left[\frac{1}{T_1} - \frac{R}{E_a} \ln \frac{\nu_1}{\nu_2} \right]^{-1} \qquad (13)$$

E_a = the potential barrier (activation energy per mole), R = the gas constant, ν_1 = original frequency, ν_2 = desired frequency, T_1 = peak temperature at 50 Hz.

This temperature frequency shift also applies to the temperature dependence of G' and $(1/\varepsilon)'$.

ACKNOWLEDGMENTS

The authors appreciate very much the collaboration of Mr. A. Heneka.

REFERENCES

1. N.G. Mc Crum, B.E. Read and G. Williams: "Anelastic and Dielectric Effects in Polymeric Solids", John Wiley and Sons; London, New York, Sidney.

2. J. Heijboer, Mechanical Properties of Glassy Polymers Containing Saturated Rings (Dissertation, University Delft, Holland) (1977)

3. P. Hedvig: "Dielectric Spectroscopy of Polymers", Adam Hilger Press, LTD Bristol (1979)

4. J.M. Schultz: "Properties of Solid Polymeric Materials", Academic Press, 1977 New York

5. W. Holzmüller, K. Altenburg: "Physik der Kunststoffe", Akademie Verlag Berlin

6. A. Messiah: "Quantum Mechanic" Vol. III, North Holland Publishing Company

7. K. Shimizu, O. Yano, Y. Wada, Dielectric Relaxations in Polymers with Pendant Phenyl Groups or Pyridine, J. Pol. Sci., Vol. 13, 2357 - 2368

8. D. Langbein: "Theory of van der Waals Attraction", Springer Tracts in Modern Physics" Heidelberg, New York.

9. W. Meyer, Messungen dielektrischer Microwellenverluste bei tiefen Temperaturen, (Dissertation, Tech. University Braunschweig)

MECHANICAL PROPERTIES OF UNIDIRECTIONALLY REINFORCED MATERIALS*

Toichi Okada and Shigehiro Nishijima

ISIR Osaka University, Ibaraki, Osaka, Japan

Hitoshi Yamaoka and Kiyomi Miyata

KURRI, Kyoto University, Kumatori, Osaka, Japan

Yasuhiro Tsuchida, Koichi Mizobuchi, and Yasuo Kuraoka

Cryogenic Technology Development Center, Hoxan Corp.
Tsukuba, Ibaraki, Japan

Shingo Namba

University of Osaka Prefecture, Sakai, Osaka, Japan

ABSTRACT

Mechanical properties of unidirectionally reinforced composite materials with the systematic variation of reinforcement and matrix have been studied aiming at standardization of materials, testing methods and developing advanced materials. Mechanical properties on identical materials were measured at three different laboratories independently and the results were compared each other.

The variation of mechanical properties originated from reinforcement, epoxy matrix and the deviation of data caused by different experimental method is discussed. The possibility of standardization and developement of advanced materials are also discussed.

INTRODUCTION

Composite materials show a variety of fracture mode which leads to the large scatter of mechanical properties. This intrinsic nature of composite materials is originated from the compositeness and/or heterogeneity of their internal structure. This is the reason why the close investigation is required in practical application of composite

---

* Work in part supported by Grant in Aid for Scientific Research No. 59050044 and No. 59750053 Ministry of Education in Japan, and Grant from the Research Project in ISIR.

materials. Furthermore, a number of components exist as reinforcement and/or matrix. Various compositing processes can be employed. Consequently, the standardization of composite materials has not been established yet, though some attempts have been made[1] (Problem 1).

The testing method of composite materials has not been established yet. Since especially fiber reinforced plastics demonstrates size effects and deformation mode and/or deformation rate dependence, it is difficult to compare materials each other or to accumulate data for design based on one certain method. The establishment of testing method has to be done urgently (Problem 2).

In the development of advanced materials, the materials should be manufactured in identical method and conditions. The measurement should be done on the specimen with identical dimension using identical method, in order to clarify the merits and demerits of the new materials compared with usual ones. In other words, the standardization of materials and testing method can be thought as the first step in the development of advanced materials. From this view-point, the standardization should be an urgent task for the development of better composite material (Problem 3).

Japanese three groups have cooperatively been dealing with the standardization of composite materials used in cryogenic temperatures. Unidirectionally reinforced composite materials were prepared with Pultrusion method[2] on the line of the standardization manufacturing process to obtain materials with same quality. The specimens were made by one manufacturer in order to minimize the scatter of properties. Concerning the testing method, each group has made the experiment under a certain specified condition which is varied systematically from group to group. The results were compared each other aiming at the establishment of better testing method. The possibility of development of advanced materials has been checked on the ceramics fiber reinforced plastics.

EXPERIMENTALS

In this work the results measured by three different groups on identical materials are shown. The mechanical test was basically three-point flexural test, and Young's modulus and breaking stress were calculated. Since each group made the experiment with different flexural speed and different jig. The effects of the flexural speed and the jig on results can be discussed.

Groups

The groups are Institute of Scientific and Industrial Research Osaka University (ISIR, including University of Osaka Prefecture), Research Reactor Institute Kyoto University (KUR) and Hoxan Cryogenic Technology Development Center (Hoxan). Each group made use of its characteristics, that is, Hoxan made the specimen, ISIR summarized this work and KUR will study the reactor irradiation effects. The testing machines used by ISIR, KUR and Hoxan are Shimazu universal testing machine RS, Shimazu Autograph universal testing machine AG-500A and Instron universal testing machine 1125, respectively (refer to Table 4).

Sample

We made it a rule to use the commercially available components. Hoxan made the sample. The samples are unidirectionally reinforced

Table 1. Makers and trade names of each reinforcement

Fiber	Maker	Trade Name	Finish	Tel(Japan)
Glass	Nitto Boseki	E-glass	silane	03-276-5282
Carbon	Toray	T-300 (high strength)	sizing	03-245-5777
Silicon Carbide	Nippon Carbon	Nicalon	sizing	03-552-6111
Alumina	Sumitomo Chemical	Sumika Alumina	----	03-278-7192

plastics made by pultrusion changing the matrix and reinforcement systematically.

(i) Components The matrix of the FRP is ARALDITE (CY205/HY905/DY040/DY061) supplied by CIBA GEIGY Co. Ltd. Two types of epoxy matrix are made changing the amount of the diluent (DY040), that is, matrix A (100/100/15/1) and matrix B (100/100/60/1) of weight fraction. The four kinds of reinforcements are chosen; they are E-glass, carbon fiber (high strength), silicon carbide fiber (SiC) and alumina (γ-Al_2O_3) fiber. Table 1 presents the maker and the trade name of each reinforcement. The properties and specifications of each reinforcement in a catalog are shown in Table 2. Hereafter glass fiber, carbon fiber, silicon carbide fiber and alumina fiber reinforced epoxy are called GFRP, CFRP, ALFRP and SFRP, respectively. CFRP is expected to show the highest specific strength. ALFRP will show good insulation. Two types of dimensions are chosen, that is, 100 mml x 3 mmϕ and 100 mml x 4 mmϕ (See Table 3). Totally 16 kinds of FRP are made, that is, two kinds of matrix, 4 sorts of reinforcements and two types of dimensions.

(ii) Manufacturing process The specimens were made by pultrusion method. The resin was impregnated with fibers in vacuum chamber. The air in the resin and among fibers was removed by vacuum pumping. After impregnation, the specimens were cured in copper tubes with

Table 2. Properties of each reinforcement from their catalog

Fiber	Young's Modulus (GPa)	Tensile Strength (GPa)	Density (g/cc)	Fiber Diameter (μm)
Glass	75.5	2.45	2.5	9
Carbon	235	2.74	1.7	8
Silicon Carbide	176	2.45	2.5	10
Alumina	206	1.76	3.2	17

Table 3. The sample dimensions of specimens

3 mmϕ Specimen	Diameter (mm)	Standard Deviation
GFRP	3.028	0.024
CFRP	3.019	0.023
ALFRP	3.023	0.024
SFRP	3.030	0.03
4 mmϕ Specimen	**Diameter (mm)**	**Standard Deviation**
GFRP	4.001	0.011
CFRP	4.001	0.014
ALFRP	4.007	0.014
SFRP	3.997	0.009

silicon release agent. Curing condition was 100 °C - 10 hours. Two sized copper tubes were prepared for two sizes of specimens. The dimensions of tubes are 220 mm in length and 3 mm or 4 mm in diameter. The fiber content of specimens were arranged 50 volume percent. After curing, the specimens were removed from copper tubes. Machining of the specimen surface was not done.

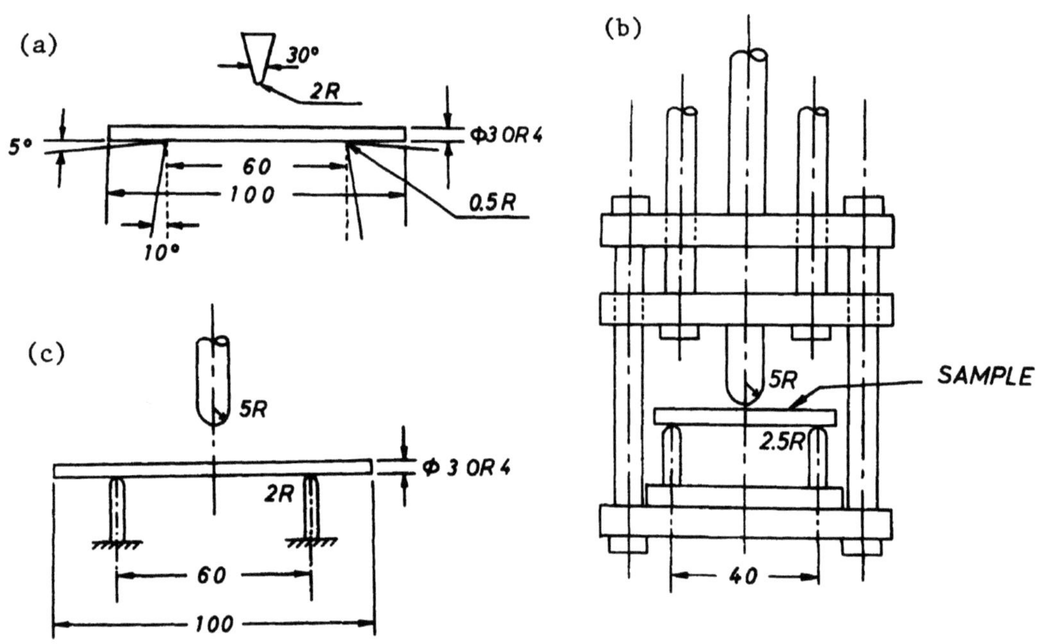

Fig. 1. Schematic illustrations of apparatus of each group; (a) ISIR (b) KUR (c) Hoxan.

Flexural Test

Schematic illustrations of jigs of each group which are designed in reference to JIS[3] and ASTM[4] are shown in Fig. 1. Rough estimation reveals that the span of the jig needs more than 40 mm, and hence the span used by ISIR and Hoxan is 60 mm and that of KUR is 40 mm. The radii of the loading tip are 2 mm (ISIR) and 5 mm (Hoxan, KUR). The radii of the supports are 0.5 mm (ISIR), 2 mm (HOxan and 2.5 mm (KUR). The deformation speeds are 1 mm/min (KUR) 1.6 mm/min (ISIR), and 5 mm/min (Hoxan). In Table 4 conditions of flexural tests of each group are demonstrated. The comparison of the results enable us to know the deformation rate dependence and tip radius dependence of the mechanical behavior. The breaking stress andYoung's modulus are calculated following formula;

$$\sigma = 8WL/\pi D^3 \tag{1}$$

$$E = 4/3 \times L^3/\pi D^4 \times F/\lambda \tag{2}$$

where σ is breaking stress, L span, W breaking force, D diameter of the specimen, and F/λ the slope of the load-displacement curve. Three specimens were tested basically at one condition.

Measurement of Static Young's Modulus

Static Young's Modulus measured on 3 mmϕ specimen. This method is basically same as three-point flexural test and the constant load is applied to the specimen and the displacement was measured. The ratio of the load and displacement is defined as F/λ in equation (1). In this case Young's modulus at small strain can be derived. The schematic illustration of the apparatus is shown in Fig. 2. The load was increased stepwise (up to 4.9 N in every 0.98 N), and the displacement of the specimen was measured by the differential transducer.

Effect of Span

ISIR and Hoxan investigated the effect of span on flexural strength cooperatively in order to analyze the data especially on GFRP because the data of GFRP was most widely scattered. The apparatus is shown in Fig. 3. The span was varied from 20 to 100 mm with step of 10 mm. The test was made on the 3 mmϕ specimen at RT and LNT.

Table 4. Experimental conditions of flexural tests of each group

	ISIR	KUR	HOXAN
Span (mm)	60	40	60
Loading tip radius (mm)	2	5	5
Supporting tip radius (mm)	0.5	2.5	2
Deformation speed (mm/min)	1.6	1	5
Testing machine	Shimazu	Shimazu	Instron

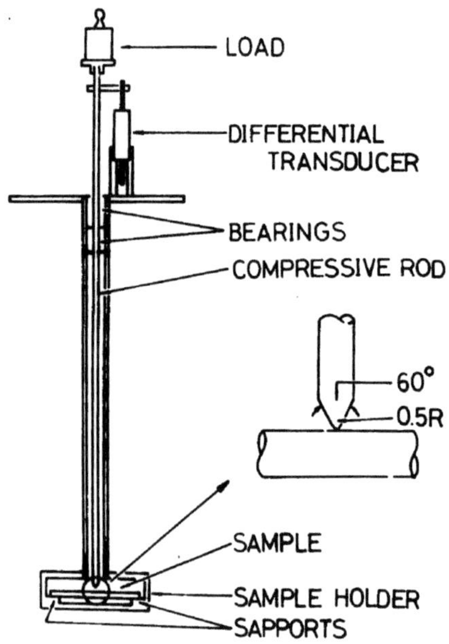

Fig. 2. Apparatus for static measurement of static Young's modulus.

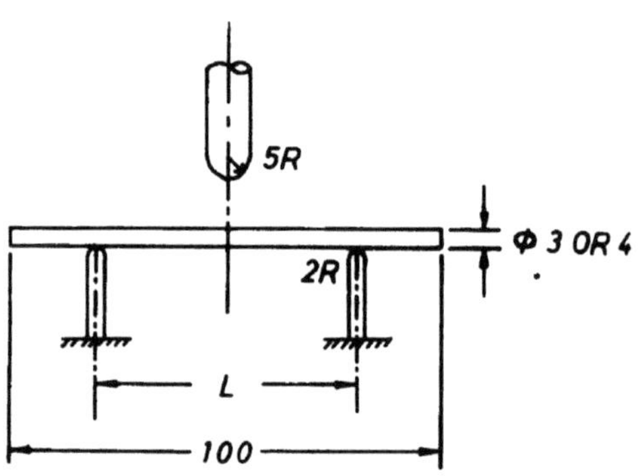

Fig. 3. Apparatus for study of span dependence of mechanical properties.

RESULT AND DISCUSSIONS

Figure 4 shows the typical load-displacement curves obtained at both RT and LNT. Young's modulus is calculated from the slope of the initial load-displacement and the breaking stress (or Yield stress) is defined as the largest stress. In Table 5, 6, 7, 8, the experimental results are presented. The values are basically the average of the three results. The number of the tested specimen is shown suffixally when it is smaller than 3. The values in parenthesis represent the standard deviation. The effects of following items on results are discussed here; materials, jig and dimension of specimen.

Comparison of Material

(i) Matrix Matrix B shows considerable plasticity at RT and hence the measurement of static Young's modulus of FRP containing matrix B could not be made. In measuring Young's modulus, the balanced displacement with the applied load should be measured though the plasticity of matrix B disturbs to define the balanced position as presented in Fig. 5. On the other hand the static Young's modulus of FRP constructed by matrix A sould be measured. Because of this static Young's modulus of FRP containing matrix B at RT are not tabulated in Table 7. This difference of matrix is reflected to the load-displacement curves. At RT the FRP containing matrix A shows linear load-displacement feature while these containing matrix B present yielding phenomena and do not show clear breaking on load-displacement

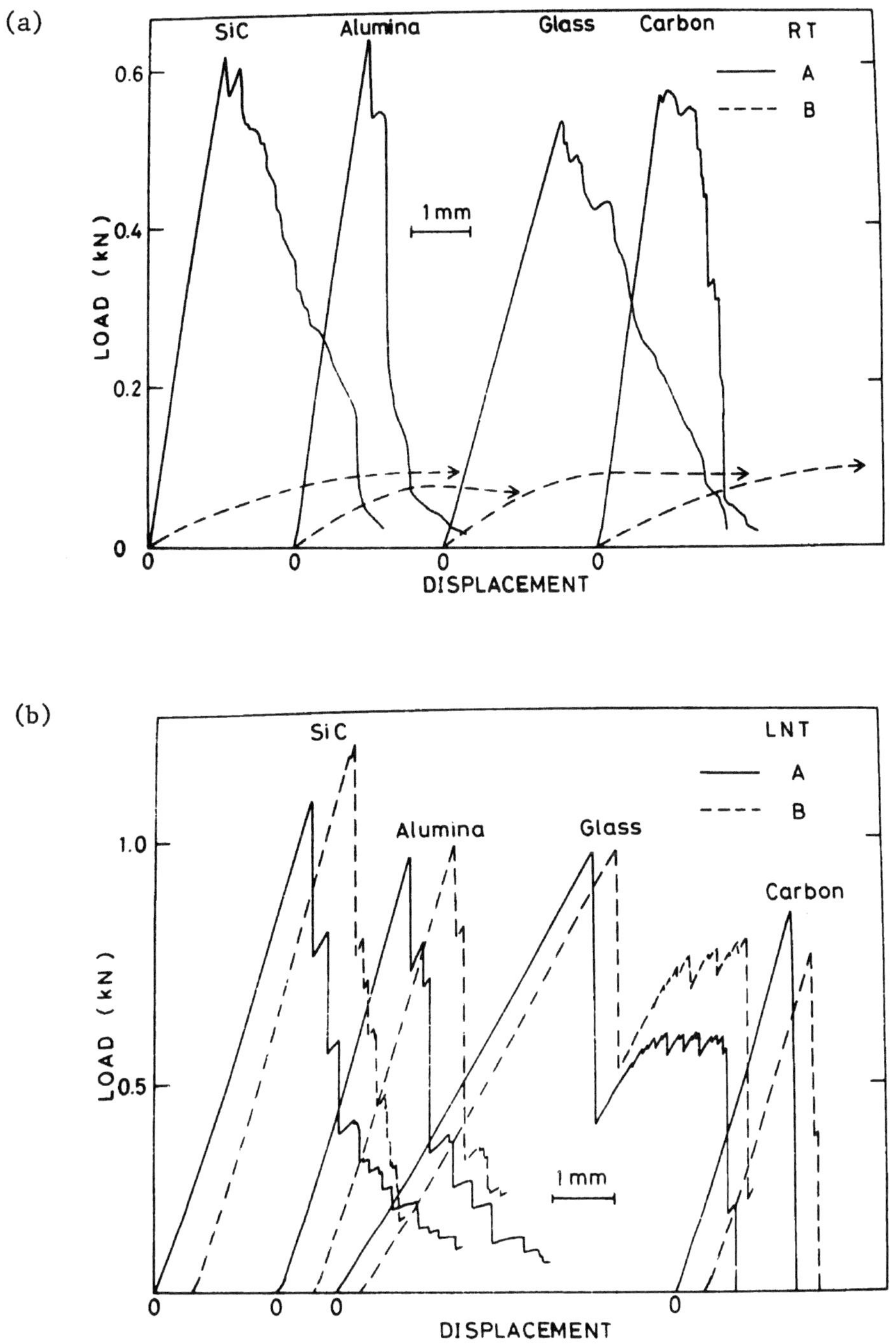

Fig. 4. Load-displacement curves of each FRP obtained at RT (a) and LNT (b)

curve. The Young's moduli of FRP containing matrix B are smaller than those containing matrix A. When the law of mixture is applied to the Young's modulus following formula is drawn.

$$E_{comp} = V_f E_f + V_e E_e \tag{3}$$

where E_{comp} is Young's modulus of composite, V_f volume fraction of fiber, V_e volume fraction of epoxy matrix, E_f Young's modulus of fiber and E_e Young's modulus of epoxy matrix. The E_e is usually very small compared with E_f and hence equation (3) can be rewritten as

$$E_{comp} = V_f E_f \tag{4}$$

Table 5. Young's modulus and breaking stress of each FRP containing matrix A of which diameter is 3 mm. The values in parenthesis present the standard deviation. The numerics presented suffixally are the number of tested specimen.

RT		GFRP	CFRP	ALFRP	SFRP
Young's Modulus (GPa)	ISIR	40.2 (3.2)	87.2 (0.59)	93.8 (1.3)	84.6 (1.7)
	KUR	33.0 (0.42)	71.9 (1.7)	73.8 (0.80)	69.8 (1.5)
	HOXAN	37.5 (0.42)	93.9 (2.7)	95.7 /2/ (0.64)	86.5 /2/ (2.2)
	Static	32.6	61.5	60.5	59.0
Breaking Stress (GPa)	ISIR	0.940 (0.017)	1.06 (0.0098)	1.19 (0.068)	1.13 (0.012)
	KUR	0.909 (0.054)	1.15 (0.056)	1.26 (0.047)	1.27 (0.023)
	HOXAN	1.19 (0.041)	1.35 (0.025)	1.46 /2/ (0.015)	1.60 /2/ (0.020)
LNT		**GFRP**	**CFRP**	**ALFRP**	**SFRP**
Young's Modulus (GPa)	ISIR	40.1 (1.4)	92.7 (2.5)	99.7 (2.3)	91.7 (2.4)
	KUR	36.7 (0.16)	77.2 (1.8)	81.2 (3.4)	76.5 (3.1)
	HOXAN	40.5 (0.88)	96.6 (0.76)	106.3 (2.3)	94.3 (1.8)
	Static	37.1	70.7	66.4	63.0
Breaking Stress (GPa)	ISIR	1.95 (0.070)	2.04 (0.30)	2.39 (0.11)	2.32 (0.087)
	KUR	2.49 (0.087)	2.07 (0.18)	2.18 (0.14)	2.31 (0.087)
	HOXAN	>2.09 (0.034)	2.00 (0.073)	2.42 (0.057)	2.19 (0.083)

In practical FRP misaligment factor, K should be multiplied and hence

$$E_{comp} = KV_fE_f \qquad (5)$$

here K is estimated to be from 0.9 ~ 1.0. Young's moduli of each FRP are calculated using (4) or (5) and the results are shown in Table 9. The Young's moduli obtained by Hoxan are shown in this table with calculated values. The Young's moduli of the FRP containing matrix B are smaller than those of calculated values though good agreements are found in those containing matrix A. This disagreement suggests the plastic deformation of matrix B. The Young's modulus should be

Table 6. Young's modulus and breaking stress of each FRP containing matrix A of which diameter is 4 mm. The values in parenthesis present the standard deviation. The numerics presented suffixally are the number of tested specimen.

RT		GFRP	CFRP	ALFRP	SFRP
Young's Modulus (GPa)	ISIR	33.6 (0.33)	72.5 (3.1)	76.7 (2.1)	74.3 (1.2)
	KUR	27.7 (0.37)	50.6 (0.95)	52.2 (0.87)	49.9 (0.97)
	HOXAN	34.6 /2/ (0.44)	80.0 (1.2)	82.3 (0.68)	87.7 /2/ (0.74)
Breaking Stress (GPa)	ISIR	0.773 (0.012)	0.921 (0.020)	1.00 (0.053)	0.990 (0.051)
	KUR	0.836 (0.017)	0.915 (0.011)	0.990 (0.024)	1.02 (0.037)
	HOXAN	0.956 /2/ (0.0034)	1.09 (0.028)	1.24 (0.049)	1.22/2/ (0.041)
LNT		**GFRP**	**CFRP**	**ALFRP**	**SFRP**
Young's Modulus (GPa)	ISIR	37.4 (1.2)	85.9 (0.78)	92.3 (1.0)	84.0 (1.0)
	KUR	34.1 (0.49)	63.6 (1.6)	68.6 (0.74)	65.7 (0.88)
	HOXAN	38.6 (0.24)	86.9 (0.91)	91.2 (2.6)	82.5 (4.5)
Breaking Stress (GPa)	ISIR	1.75 (0.066)	1.70 (0.087)	2.15 (0.054)	2.36 (0.041)
	KUR	2.10 (0.023)	1.84 (0.046)	2.13 (0.050)	2.40 (0.034)
	HOXAN	2.10 (0.068)	1.84 (0.070)	2.18 (0.070)	2.34 (0.25)

defined at the region where plasticity is negligibly small, that is, as the initial tangent of load-deflection curve. The large deformation rate dependence of mechanical properties on the FRP containing matrix B is expected. The static Young's modulus should be smaller than these obtained by flexual test. On the other hand at LNT the difference of Young's moduli originates from difference in matrix come to be small. This means that the plastic deformation is frozen to some extent even in matrix B and the equation (5) is supported.

(ii) Reinforcement Concerning breaking stressess, CFRP shows the lowest value both at RT and LNT while the carbon fiber itself shows the highest value among the reinforcements. This appears to reflect the fact that the flexural strength of carbon fiber is relatively low compared with the tensile strength. The breaking stress of ALFRP and

Table 7. Young's modulus, yield stress or breaking stress of each FRP containing matrix B of which diameter is 3 mm. The values in parenthesis present the standard deviation. The numerics presented suffixially are the number of tested specimen.

RT		GFRP	CFRP	ALFRP	SFRP
	ISIR	20.6	41.2	40.6	40.7
Young's		(1.7)	(5.3)	(7.6)	(5.3)
Modulus	KUR	7.54	9.71	8.15	7.36
(GPa)		(0.34)	(1.1)	(2.6)	(1.0)
	HOXAN	29.7 /1/	64.8	65.9 /2/	64.1 /2/
		(0.0)	(3.6)	(0.69)	(1.1)
	ISIR	0.192	0.285	0.248	0.268
Yield		(0.012)	(0.0069)	(0.0070)	(0.0095)
Stress	KUR	0.124	0.161	0.127	0.124
(GPa)		(0.0046)	(0.0059)	(0.013)	(0.016)
	HOXAN	0.333 /1/	0.357	0.403 /2/	0.418 /2/
		(0.0)	(0.027)	(0.020)	(0.014)

LNT		GFRP	CFRP	ALFRP	SFRP
	ISIR	38.2	96.1	102.3	92.8
Young's		(1.7)	(0.74)	(5.6)	(4.0)
Modulus	KUR	36.0	79.5	82.9	73.1
(GPa)		(0.13)	(2.9)	(1.2)	(3.7)
	HOXAN	38.6 /2/	99.0	103.5	97.9 /2/
		(0.44)	(2.7)	(5.5)	(2.9)
	Static	30.4	60.3	----	66.7
	ISIR	2.12	1.94	2.40	2.43
Breaking		(0.16)	(0.057)	(0.095)	(0.15)
Stress	KUR	2.51	1.97	2.33	2.35
(GPa)		(0.18)	(0.055)	(0.069)	(0.14)
	HOXAN	1.93 /2/	2.12	2.32	2.39 /2/
		(0.15)	(0.070)	(0.11)	(0.12)

SFRP are almost same at LNT. Concerning the GFRP the close discussion will be made. The FRP containing matrix B does not show apparent fracture at RT and hence the yield stresses instead of breaking stress are shown. The yield stresses are almost constant among different materials in each group. The mechanical properties of the FRP constructed by matrix B is dominated by the matrix instead of reinforcement at RT. This feature have close correlation to the deformation mode. In flexural deformation the matrix should share the stress considerably as interlaminar shear stress. This is the reason that the mechanical properties of the FRP containing matrix B is dominated by the matrix at RT. Generally speaking the Young's moduli of FRP

Table 8. Young's modulus, yield stress or breaking stress of each FRP containing matrix B of which diameter is 4 mm. The values in parenthesis present the standard deviation. The numerics presented suffixally are the number of tested specimen.

RT		GFRP	CFRP	ALFRP	SFRP
Young's Modulus (GPa)	ISIR	19.9 (0.26)	32.8 /2/ (2.4)	33.2 (3.9)	40.2 /2/ (0.31)
	KUR	5.18 (0.35)	6.81 (2.4)	6.08 (1.4)	4.62 (0.92)
	HOXAN	27.6 (0.32)	51.9 /2/ (1.4)	48.4 /2/ (2.9)	50.9 /2/ (0.34)
Yield Stress (GPa)	ISIR	0.224 (0.0050)	0.284 /2/ (0.0039)	0.212 (0.024)	0.249 /2/ (0.0093)
	KUR	0.134 (0.082)	0.181 (0.014)	0.138 (0.021)	0.139 (0.0077)
	HOXAN	0.324 (0.014)	0.389 /2/ (0.0098)	0.309 /2/ (0.059)	0.360 /2/ (0.0083)
LNT		**GFRP**	**CFRP**	**ALFRP**	**SFRP**
Young's Modulus (GPa)	ISIR	35.3 (1.0)	74.3 /2/ (0.19)	85.2 (5.0)	85.1 /2/ (3.1)
	KUR	34.3 (0.42)	63.4 (0.0)	65.3 (0.88)	67.8 (2.7)
	HOXAN	38.8 (0.98)	86.7 /2/ (0.29)	94.1 (1.5)	92.0 /2/ (2.8)
	Static	30.4	60.3	----	66.7
Breaking Stress (GPa)	ISIR	1.88 (0.053)	1.69 /2/ (0.12)	2.21 (0.081)	2.53 /2/ (0.063)
	KUR	2.17 (0.021)	1.70 (0.091)	2.12 (0.063)	2.55 (0.079)
	HOXAN	2.19 (0.0056)	1.85 /2/ (0.0098)	2.23 (0.066)	2.38 /2/ (0.22)

show the following tendency except those obtained at RT in matrix B,

$$E_{AL} > E_C = E_S > E_G$$

where E_{AL} is Young's modulus of AFRP, E_C that of CFRP, E_S that of SFRP and E_G that of GFRP. The CFRP does not agree with the estimation from the fiber itself.

Influence of Jig

Concerning breaking stress, the data obtained at ISIR are smaller than those at KUR and Hoxan. The smaller radii of loading rod causes

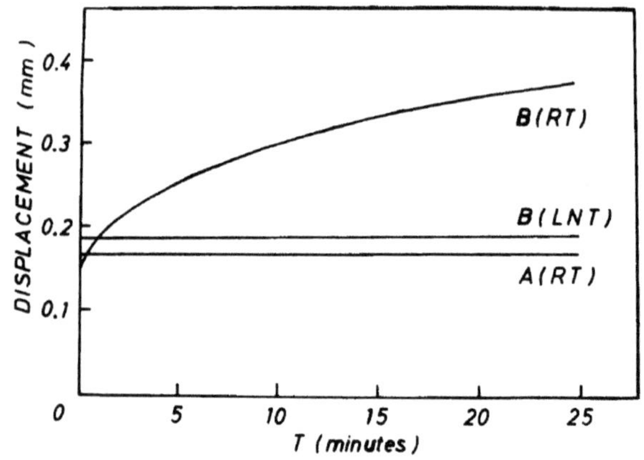

Fig. 5. Displacement-time curves in measuring static Young's modulus.

the compressive stress concentration just under the loading rod. Since the compressive strength in unidirectionally reinforced plastics is smaller than the tensile strength, just under the loading rod the concentrated compressive stress introduces the cracks easily and hence the fracture occurs at the lower load level. The four point flexural test, therefore, would be preferable.

At LNT the strength of all materials except GFRP coincides each other within experimental error (concerning GFRP close discussion will be made later which suggests the stress concentration effect is decreased at low temperature. The matrix come to be harder and increase the stiffness as temperature decreases which would lead to the decentralization of the stress. To eliminate the effect of compressive concentrated stress, the tip radius should be larger than 2 mm within the scope of present experiment.

The Young's moduli obtained at ISIR and Hoxan coincide though those at KUR are smaller than those at other laboratories. This discrepancy would not be caused by the difference of deformation rate, for strain rate at KUR is almost same at ISIR. The difference of span does not cause such large disagreement either. If the span causes the disagreement, the breaking stress obtained at KUR should not coincide to those obtained at Hoxan. The main cause of the discrepancy would be the stiffness of testing machine. The jig used at KUR was designed for lower load level. In FRP experiment, the stiffness of machine is very important especially at low temperature because of following two reasons; (i) Young's modulus increases markedly at the low temperatures compared with those at RT. (ii) The apparatus has the tendency of increase in its length and decrease in its cross

Table 9. Young's modulus obtained by calculation and experiments

Matrix	E (Theory) (GPa)	E (Matrix A) (GPa)	E (Matrix B) (GPa)
GFRP	34.0 - 37.8	37.5	29.7
CFRP	105.8 -117.5	93.9	64.8
ALFRP	79.2 - 88	95.7	65.9
SFRP	92.7 -103	86.5	64.1

section in order to reduce the thermal conductivity for low temperature use.

Influence of Sample

The breaking stress obtained on the materials of which diameter is 3 mm and 4 mm coincides each other within experimental error. Young's moduli obtained with 3 mmϕ specimen are larger than those with 4 mmϕ specimen. This origin has not been clarified yet.

Close Consideration on GFRP

The GFRP is closely studied because the scatter of data is comparatively large among groups. Figure 6 and 7 show the span length dependence of Young's modulus and breaking stress obtained on 3 mmϕ GFRP specimen at RT and LNT. At RT (Fig. 6) the breaking stress and

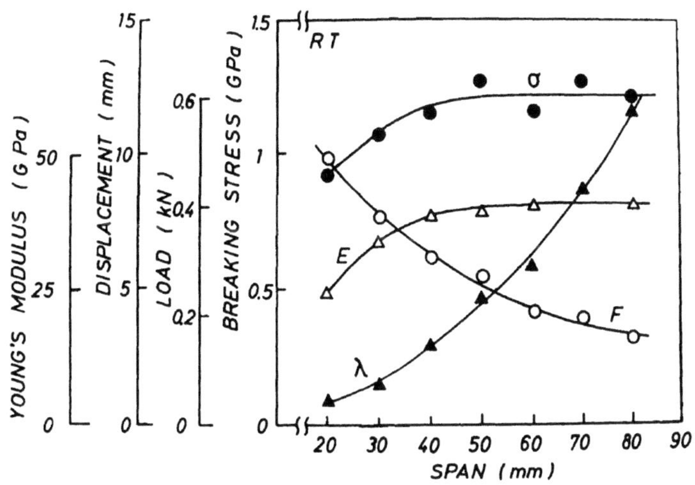

Fig. 6. Effect of span on breaking load (σ), breaking load (F), breaking displacement (λ), and Young's modulus (E) obtained at RT.

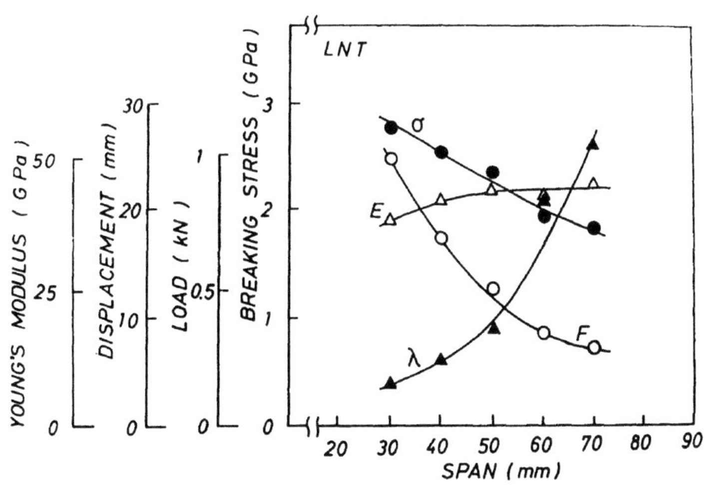

Fig. 7. Effect of span on breaking load (σ), breaking load (F), breaking displacement (λ), and Young's modulus (E) obtained at LNT.

Young's modulus increase at first and then saturate as the span increases. The cause of the low breaking stress at small span should be the coexistence of shear fracture. The maximum tensile stress σ and maximum shear stress τ are calculated as follows.

$$\sigma = 8PL/\pi D^3$$

$$\tau = 16P/3\pi D^2$$

where P is applied load, D diameter, L span. The applied load which causes the tensile fracture, P_σ and that causes the shear fracture P_τ are defined as follows.

$$P_\sigma = \pi D^3 \sigma_0/8L$$

$$P_\tau = 3\pi D^2 \tau_0/16$$

where the σ_0 is tensile strength and τ_0 shear strength. When $P_\tau > P_\sigma$, tensile fracture occurs, while shear fracture occurs under $P_\tau < P_\sigma$. Tensile fracture occurs if

$$P_\tau > P_\sigma \text{ that is } L/D > 2\sigma_0/3\tau_0 \tag{6}$$

Table 10. Change of λ/L with span

Span (L:mm)	Displacement (λ:mm)	λ/L
50	4.6	0.092
60	5.8	0.097
70	8.6	0.12
80	11.5	0.14

Table 11. Relationship between λ/L and corrected breaking stress

Span (L:mm)	Displacement (λ:mm)	λ/L	Breaking Stress (σ:GPa)	Corrected Breaking Stress $\sigma(1+4(\lambda/L)^2)$
30	4	0.13	2.70	---
40	6	0.15	2.50	2.72
50	9	0.18	2.30	2.60
60	20.5	0.34	1.89	2.77
70	26	0.37	1.76	2.73

Shear fracture occurs if

$$P_\tau < P_\sigma \text{ that is } L/D < 2\sigma_0/3\tau_0 \qquad (7)$$

When the L/D is small, the shear fracture is predominant compared with tensile fracture. Rough estimation was made using equation (6) or (7). Here let D = 3 mm, σ_0 = 1200 MPa, and τ_0 = 80 MPa at RT, then the tensile fracture region is calculated as L/D > 10 that is L should be larger than 30 mm. At LNT let σ_0 = 2000 MPa and τ_0 = 100 MPa, then we obtain L > 40 mm. Figure 6 reveals the constant values of breaking stress and Young's modulus at the span of larger than 40 mm. This is the reason why KUR chose the 40 mm span.

At LNT the increase of span decreases the tensile strength (Fig. 7). This appears to be brought by the fact that at larger region of λ/L the simple beam theory cannot be accepted, where λ is displacement of the center of the specimen. The acceptable value of λ/L are calculated from the results obtained at RT on constant breaking stress renge, that is, at the L range from 50 to 80 mm. The results are shown in Table 10. In this work, λ/L value appears to be acceptable up to 0.14, based on the results of Table 10. This value is applied to the results obtained at LNT. The results are presented in Table 11. At the larger span than 40 mm the value of λ/L is too large to use the simple beam theory. It is necessary to make a few correction at large λ/L value.[3] The correction factor is introduced in these results and the corrected results are also presented in Table 11. It is found that good agreement is obtained even at large λ/L region. It is necessary to make correction of flexural strength at large λ/L region (larger than 0.14).

CONCLUSIONS

Japanese three laboratories have studied the mechanical properties of unidirectionally reinforced plastics aiming at standardization of material, testing method and development of advanced materials. Following conclusions were drawn.

(1) Standardization of material

i) The materials of which mechanical properties are thought to be uniform could be manufactured easily by pultrusion method though there are still some problems to be solved.

ii) Since these materials are reinforced unidirectionally, they show large anisotropy and hence it is difficult to perform precise experiment. In flexural test the materials have tendency to show large deformation in which simple beam theory could not be accepted.

iii) Pultrusion method is the easy way to manufacture the specimen for the screening test of materials.

(2) Testing Method

i) The radius of the loading tip should be large enough not to cause the stress concentration.

ii) The span in flexural test should be within a certain range in order to give rise to tensile fracture and not to cause large deformation.

iii) The four point flexural test is preferred to three point test because in three point test just under the loading tip the compressive fracture could be introduced which results in inaccuracy of obtained data.

iv) The stiffness of the testing machine is very important to calculate Young's modulus. Since the initial tangent is needed to calculate the accurate Young's modulus, the design of cryostat should be made carefully not to decrease the stiffness of testing machine.

(3) Development of advanced materials

i) The ALFRP show the largest Young's modulus followed by CFRP and SFRP which present almost same Young's modulus. The Young's modulus of ceramics fiber reinforced plastics are more than double of that of GFRP both at RT and LNT.

ii) Breaking stresses of ALFRP and SFRP are almost same at low temperature. Breaking stress of CFRP obtained flexural test is small considerably compared with the expected value from tensile strength.

iii) GFRP can show large deformation at low temperature. This is the characteristics of GFRP.

iv) Ceramics fiber reinforced plastics would be promissing materials as cryogenic materials.

ACKNOWLEDGEMENTS

The authors are grateful to the menbers of Okad Laboratory, especially to Mr. M. Takeno, Mr. K. Takahata and Miss T. Ichida for their help.

REFERENCES

1. M. B. Kasen, Standardizing nonmetallic composite materials for cryogenic applications, in: "Nonmetallic Materials and Composites at Low Temperatures 2", G. Hartwig and D. Evans, eds., Plenum Press, New York (1982), pp. 327-337.
2. M. B. Kasen, Method for Making Fiber-Reinforced Specimens for Use in Screening of Cryogenic Radiation Performance, CEC-ICMC Colorado Springs (1983) CP-13.
3. JIS K 7203-1977, Flexural Test of Regid Plastics.
4. ASTM D 790-71, Standard Methods of Test for Flexural Properties of Plastics.

DYNAMIC YOUNG'S MODULUS AND INTERNAL FRICTION IN COMPOSITE MATERIALS*

Shigehiro Nishijima, Kenichi Matsushita, Toichi Okada, and Taira Okamoto

Institute of Scientific and Industrial Research
Osaka University
Ibaraki, Osaka, Japan

Takeshi Hagihara

Osaka Kyoiku University

ABSTRACT

The strain amplitude dependence of Dynamic Young's modulus and internal friction have been studied on variously fatigued glass cloth reinforced epoxies and was compared with laser transmittance. The temperature dependence was also measured in temperature range from 100 to 380 K.

The strain amplitude dependence in Young's modulus and internal friction come to notable as the fatigue progresses. The close correlation was revealed between laser transmittance and strain amplitude dependence. The degree of interface fracture could be estimated by strain amplitude dependence. The different temperature dependence of internal friction were found in different FRP of which Young's modulus are just same at room temperature, because the temperature dependence of internal friction mainly reflects the behavior of matrix in FRP.

INTRODUCTION

The fracture mode of composite materials are complicated and varied originated from their heterogenity. Among them, the fracture of composite materials often starts with interface failure. Such failures degradate the over-all mechanical properties of composite materials considerablly even though the reinforcements were not fractured. It is, therefore, important to detect these failures before the over-all fracture in practical application of composite materials.

We have reported that the interface failure could be detected by means of dynamic Young's modulus and internal friction.[1] In this work translucent GFRP (Glass Reinforced Plastics) was employed and various

---

* Work in part supported by Grant in Aid for Scientific Research No. 59050044 and No. 59750053 Ministry of Education in Japan, and Grant from the Research Project in ISIR.

degree of failure are introduced by fatigue test in order to make the interlaminar failure visible. The Young's modulus and internal friction behavior of GFRP are correlated with the laser transmittance. A model which explains the amplitude dependence of Young's modulus and internal friction is given. It is also reported that the GFRP containing different matrix could be distinguished by means of temperature dependence of internal friction.

EXPERIMENTALS

A fatigue test was performed on commercially available composite materials. Dynamic Young's modulus and internal friction were measured on variously fatigued specimen. The laser transmittance was also measured and were compared with the results on Young's modulus and internal friction. The temperature dependence of internal friction were measured on different GFRP and the results were compared.

Specimen

The materials tested are commercially available glass cloth reinforced epoxies supplied by Hoxan Co. (named A) and Nitto Electric Industrial Co. (called B). These are defined as E1F45P-E/76(A) and E1F42P-E/52(B), respectively.[2] The specimen 100 mm x 10 mm x 2 mm were cut from a sheet of 2 mm thickness in the warp direction. Sample A is translucent relatively and hence the laser transmittance measurement was made on sample A.

Fatigue Test

Reversed flexural stress (4 point bending : R = -1) was applied to the specimen at the frequency of 30 Hz at room temperature. The temperature of the specimen was measured with a thermocouple attached on the surface during the fatigue test. The fatigue test was performed at the 70% of static breaking stress and the degree of the failure was classified tentatively by the temperature change of the specimen.[3,4] The dimensions and supporting system is shown in Fig. 1.

Dynamic Young's Modulus and Internal Friction

Dynamic Young's modulus and internal friction were measured on variously fatigued specimens. The flexural vibration and the free-decay method were used for measurements of Young's modulus and internal frictions, respectively. The measurements were performed on the strain amplitude range from 10^{-10} to 7×10^{-7} at room temperature

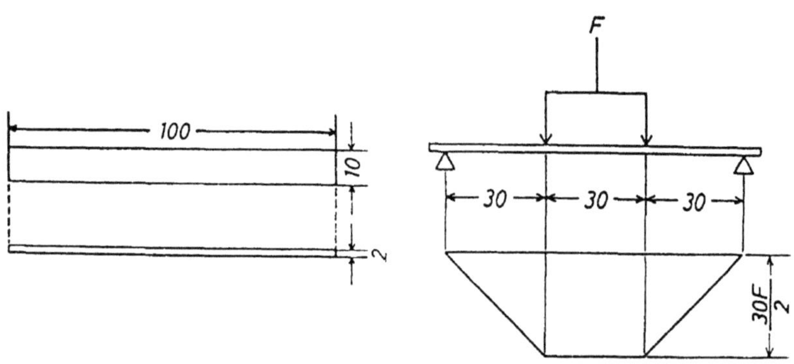

Fig. 1. Dimensions and supporting system of specimens. In this case moment is defined as 30F/2 where F is applied force.

(RT; 295 K) or down to 130 K on the fixed strain amplitude of approximately 7×10^{-7}. The changing rate of temperature was 3 K/min. The dynamic Young's modulus (kg/mm^2) was calculated from the following formula;

$$E = 9.467 \times 10^{-7}\ l^3 w f^2/d^3 b \qquad (1)$$

where l is the length of the specimen (cm), w the weight of the specimen (g), f resonant frequency, d thickness (cm) and b width (cm). The resonant frequency was in the range of 600-800 Hz.

Internal friction was derived from the logarithmic decrement using the following formula;

$$Q^{-1} = -1/(\pi f t) \times \log(A/A_0) \qquad (2)$$

where A is strain amplitude, A_0 initial strain amplitude and t is time after the swiching off. The derivation of internal friction was shown in Fig. 2. The strain amplitude dependence was measured by changing the initial strain amplitude, A_0. The strain amplitude was calibrated by strain gauge and calculation from the deviation of the center of the specimen. A good agreement between observed and calculated values was obtained.

Laser Attenuation Measurement

The laser transmittance of the specimen was measured using He-Ne laser on fatigued translucent specimen(A). The laser attenuation coefficient, λ is defined as following formula;

$$I = I_0 \exp(-\lambda d) \qquad (3)$$

where I is intensity of the transmitted light, I_0 initial intensity of light, and d thickness of the specimen (cm). The interlaminar failure would reduce the laser transmittance results in increase of laser attenuation coefficient. The correlation between laser attenuation and dynamic Young's modulus and/or internal friction was examined.

RESULTS AND DISCUSSIONS

The six variously fatigued specimen were prepared. The tentative rough determination of the fatigue degree was made by the temperature

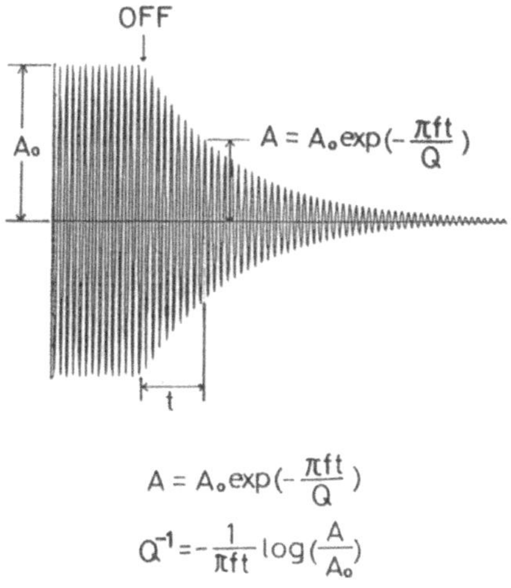

Fig. 2. Derivation of internal friction. The internal friction is calculated from the logarithmin decrement of the strain amplitude.

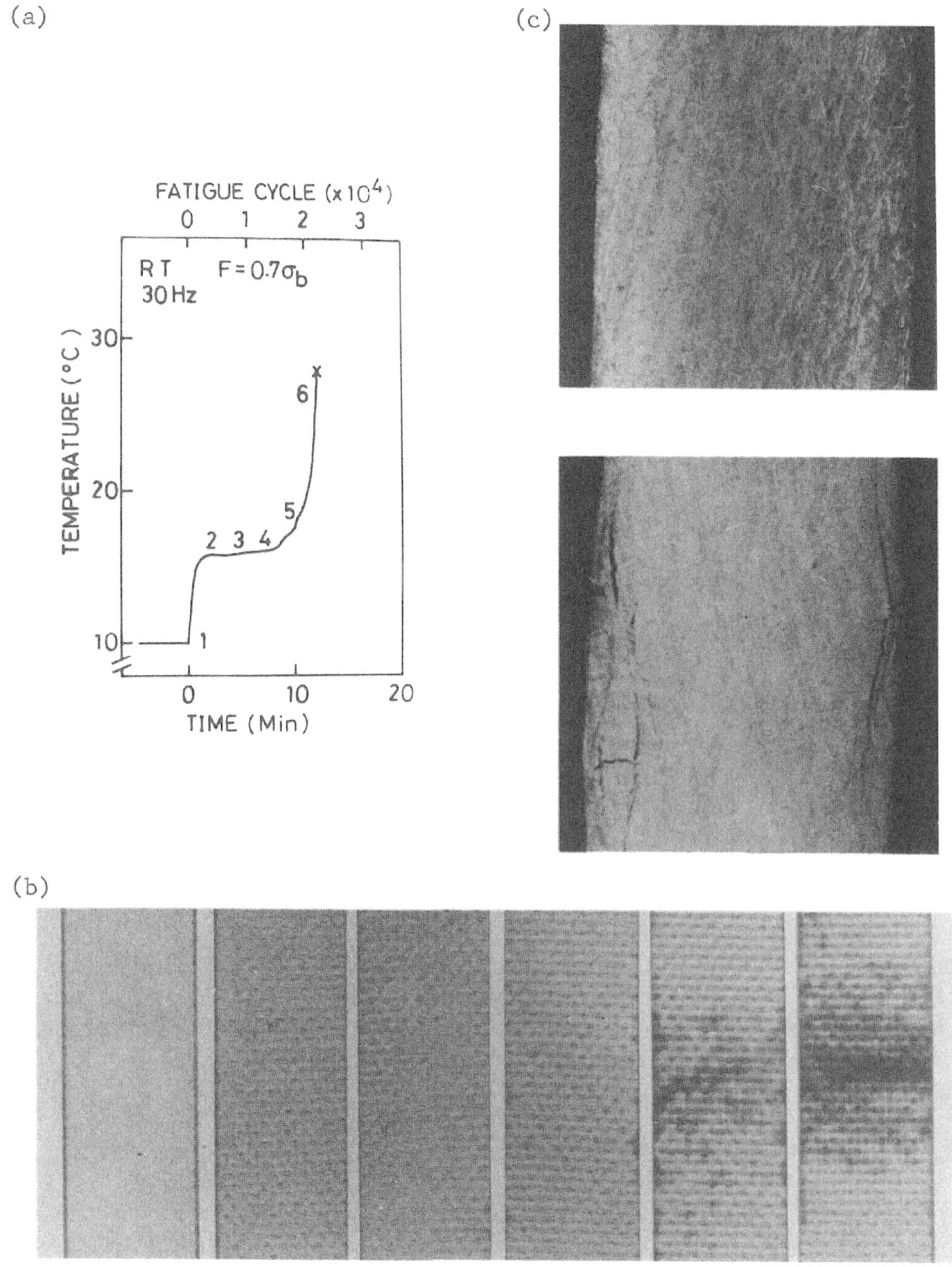

Fig. 3. (a) Temperature change of specimen during fatigue test. Each number of the temperature stage corresponds to that of the specimen. (b) Photogragph of fatigued specimen. (c) Side view of virgin (upper) and fatigued specimen (bottom).

change during the fatigue test. The temperature of the specimen generally shows sudden increase at the beginning of the test and just before the fracture as presented in Fig. 3 (a). The fatigue test was stopped at each temperature stage (No.1 means virgin sample). Figure 3 (b) shows the photographs of the fatigued specimen. The interlaminar failure is demonstrated as dark area in this photograph. In this figure the sample numbers correspond to those of temperature stage. Hereafter the sample numbers are defined as those in this figure. The side view of No. 1 and No. 6 are presented in Fig. 3 (c). The interlaminar failure can be seen on fatigued specimen.

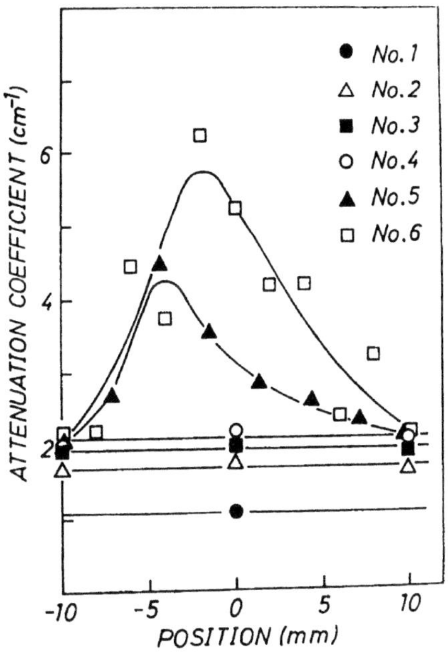

Fig. 4. Laser attenuation in fatigued specimen. Each number corresponds to that in Fig. 3 (a).

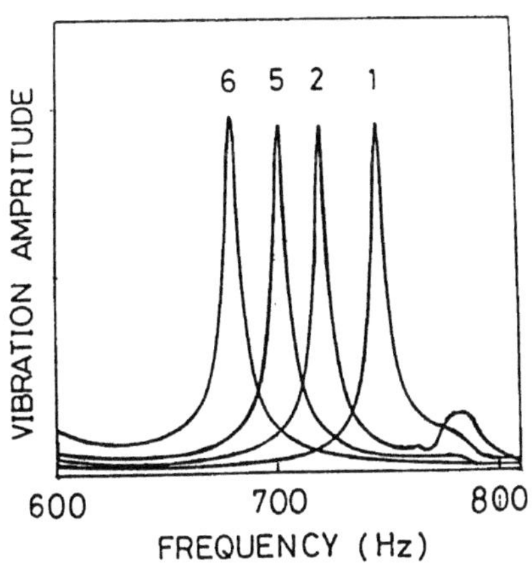

Fig. 5. Resonant curves of variously fatigued specimen obtained at RT.

Figure 4 shows the laser attenuation coefficient on each specimen. The abscissa shows the position of the specimen (i.e. the zero position means the center of the specimen). The ordinate shows the attenuation coefficient λ. The increase of the number (from 1 to 6) means the increase of the fatigue degree and hence the increase of the fatigue degree and hence the increase of fatigue degree brings the increase of the attenuation as expected from Fig. 3 (b). The interface failure induces the random reflection of light and then the attenuation coefficient of laser light increases.

Figure 5 shows the resonant curves obtained at RT on fatigued specimen. The abscissa and the ordinate present frequency and vibration amplitude of the specimen, respectively. The numbers in this figure correspond to those in Fig. 3 (b). The increase of fatigue degree induces the decrease of Young's modulus (equation (1)).

The strain amplitude dependence of Young's modulus and internal friction are presented in Fig. 6. The number in this figure means the sample number as shown in Fig. 3 (b). The Young's modulus decreases with progress of fatigue as expected in Fig 5. The amplitude dependence of internal friction comes to be notable as the

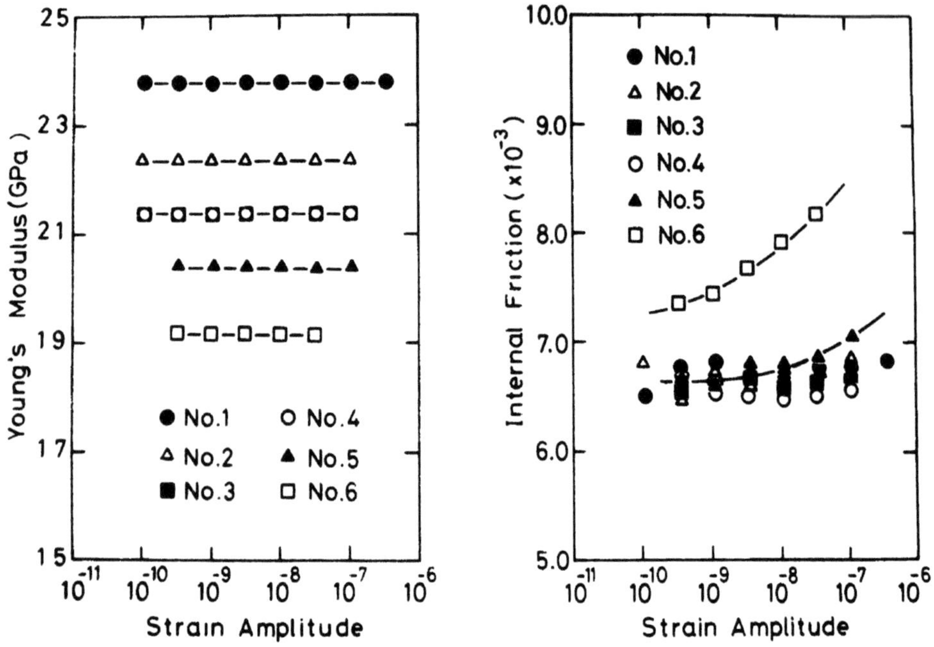

Fig. 6. Strain amplitude dependence of Young's modulus and internal friction obtained at RT.

fatigue degree increases. The internal friction shows rapid increase with strain amplitude and the strain amplitude at which the internal friction shows rapid increase decreases as the fatigue progresses. This strain amplitude coreespond to the beginning of Young's modulus decrease which is not notable in this figure. Figure 7 shows the more details of change of Young's modulus and internal friction on typical specimen. The fracture degree of interface could be estimated by the decrease of Young's modulus and amplitude dependence of internal friction. This features agree well to the results of laser attenuation. It is confirmed that the degree of interlaminar failure could be estimated by dynamic Young's modulus and internal friction.

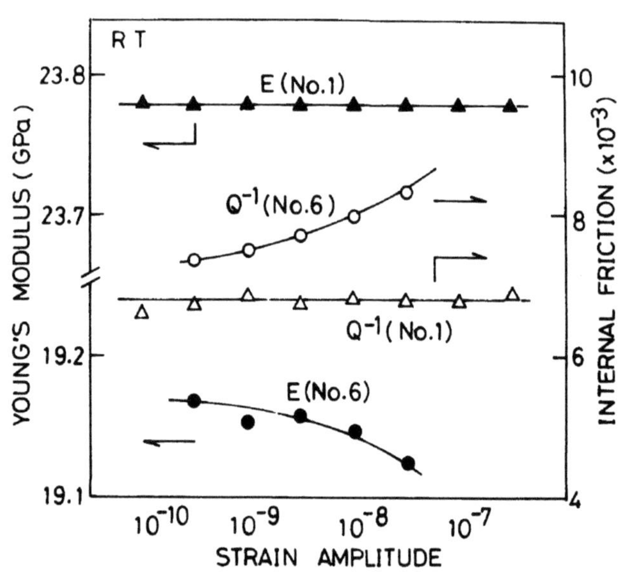

Fig. 7. Relative change of Young's modulus and internal friction with strain amplitude on virgin and fatigued specimen.

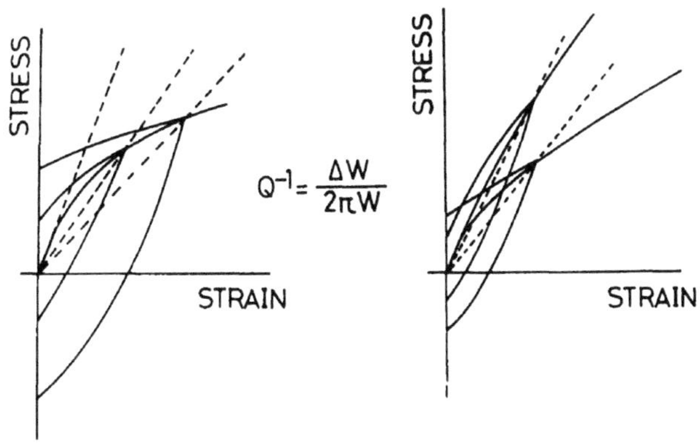

Fig. 8. A model of strain amplitude dependence of dynamic Young's modulus and internal friction.

A model of the strain amplitude dependence of Young's modulus and internal friction is presented in Fig. 8. The stress-strain curves of fatigued specimen would bend due to interface failure. The internal friction is calculated as the area of hysteresis loop of stress-strain curves and the dynamic Young's modulus is presented as the average slope of stree-strain curve (the slope of the breaking line). The increase of strain amplitude brings the larger bending of stress-strain curve and results in larger hysteresis loop and smaller Young's modulus. The drastic change of them is brought when the stress-strain curve deviate from the initial straight line and this is thought to correspond to the beginning of slipping at the interface. The stress-strain curve of more severely fatigued specimen would bend more markedly and hence even at the identical strain amplitude, the more severely damaged specimen shows larger internal friction and smaller Young's modulus.

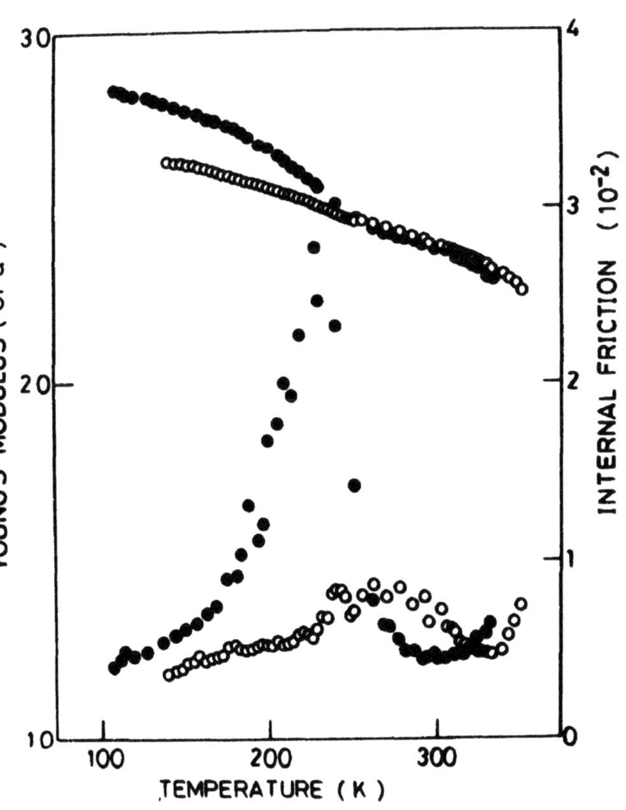

Fig. 9. Temperature dependence of Young's modulus and internal friction in different GFRP. (A:open circles, B;closed circles)

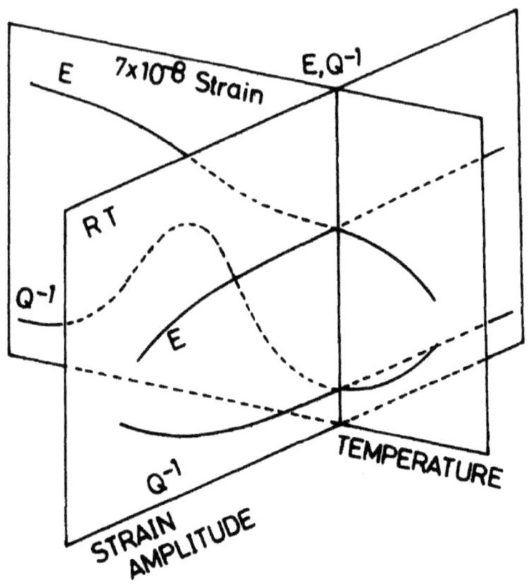

Fig. 10. Schematic illustration of temperature and strain amplitude dependence of Young's modulus and internal friction

Temperature Dependence of Internal Friction and Young's Modulus

The temperature dependence of internal friction and Young's modulus in both virgin specimen (A:open circles, B:closed circles) are presented in Fig. 9. Both GFRP show almost same Young's modulus at RT because both contain almost same fraction of glass fiber. The temperature dependence of internal friction is different each others as demonstrated in Fig. 8. This internal friction is thought to be originated from epoxy matrix[5,6] and hence the different matrix could be distinguished in terms of temperature dependence of internal friction.

Because of "ΔM effect", the temperature dependence of Young's modulus are different each other caused by the different temperature dependence of internal friction. This could mean that the change or degradation of matrix could be detected in terms of temperature dependence of internal friction. The schematic illustration of temperature and strain amplitude dependence of Young's modulus and internal friction is demonstrated in Fig. 10. Temperature dependence reveals the matrix change or degradation and strain amplitude enables the estimation of the interlaminar failure.

CONCLUSIONS

The strain amplitude dependence of dynamic Young's modulus and internal friction on the fatigued composite material were measured and were compared with laser attenuation. The temperature dependence of internal friction were also measured in different GFRP and following conclusions were drawn.
(1) The Young's modulus decreases and the internal friction increases as fatigue progresses. The results agree well to the laser attenuation and show the degree of interface failure.
(2) The rapid increase of internal friction and decrease of Young's modulus against.strain amplitude have close connection each others and this strain amplitude is thought to correspond to the beginning of slipping at the interface.
(3) The different matrix could be distinguished by means of temperature dependence of internal friction even in the FGRP which shows the almost same Young's modulus at RT.

REFERENCES

1. T. Okada, S. Nishijima, K. Matsushita, T. Okamoto, H. Yamaoka and K. Miyata, Dynamic Young's modulus and internal friction in composite material at low temperature, in: "Advances in Cryogenic Engineering," vol. 30, Plenum Press, New York (1984), pp. 9-16.
2. M. B. Kasen, Standardizing nonmetallic composite materials for cryogenic applications, in: "Nonmetallic Materials and Composites at Low Temperatures 2", G. Hartwig and D. Evans, eds., Plenum Press, New York (1982), pp. 327-337.
3. J. J. Nevadunsky, J. J. Jucas and M. J. Salkind, Early fatigue damage detection in composite materials, J. Comp. Mat. 9:394-408 (1975).
4. L. Nielsen, Fatigue Behavior of Some Filled Polymers, J. Comp. Mat. 9:149-156 (1975).
5. T. Hirai and D. E. Kline, Dynamic mechanical properties of nonstoichiometric amin-cured epoxy resin, J. Appl. Polym. Sci. 16:3145-3157 (1972).
6. T. Hirai and D. E. Kline, Dynamic mechanical properties of graphite-epoxy and carbon-epoxy composite, J. Comp. Mater. 7:160-177 (1973).

LOW TEMPERATURE DUCTILE MATRICES FOR ADVANCED FIBER COMPOSITES

G. Hartwig

Nuclear Research Center Karlsruhe, IRB/ZSM

Karlsruhe, Federal Republic of Germany

INTRODUCTION

Fiber composites are attractive alternatives to metals because of their high specific strength or stiffness or their excellent fatigue behavior. They are a necessary supplement to metals because of their low electrical and thermal conductivities, the latter being related to strength or stiffness[1]. Their disadvantage arises from the weak polymeric matrix and results in a low interlaminar shear strength and a low transverse strength. At low temperatures the majority of currently used matrices are brittle and they do not permit relaxation of residual stresses and stress concentrations to occur. In the course of cooling the different thermal contractions of fiber and matrix give rise to thermal residual stresses and strains which influence most of the mechanical properties.[2]

There are some thermoplastic polymers, such as polycarbonate (PC), polysulfone (PSU), polyethersulfone (PES) and polyethylene (HDPE) which show some low temperature ductility even at 4.2 K. Their use as a matrix for fiber composites in cryogenic technology is discussed. Problems arising from manufacturing and fiber-matrix adhesion are treated in the following papers.[3]

FREE FRACTURE STRAIN ε^{f}_{M} OF A MATRIX IN FIBER COMPOSITES

The main reasons for matrix failures are brittleness and thermal residual stresses. With decreasing temperature the polymeric matrix becomes stiffer and stronger but also less ductile, and in combination with low contraction of the fibers more and more thermal residual tensile stress and strain ε^{R}_{M} is accumulated on the matrix. This reduces the effective free stress or strain ε^{f}_{M} available for external loading.

The free fracture strain is determined by:

- the fracture strain of the matrix ε_{UTM}
- the relative thermal contraction of fiber and matrix
- the creep relaxation during cool down.

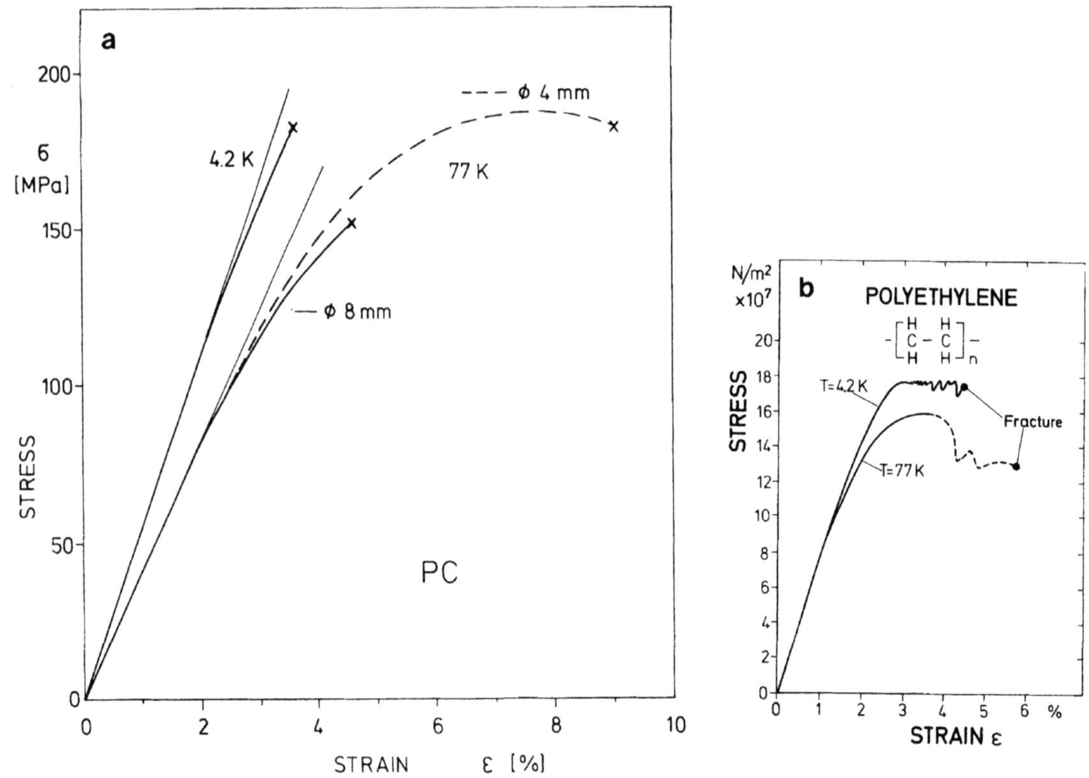

Fig. 1 Stress-strain diagrams of a) polycarbonate and b) polyethylene

Both the latter determine the thermal residual strain ε_M^R.

It holds:

$$\{\varepsilon_M^f\} = \varepsilon_{UTM} - \{\varepsilon_M^R\} \quad (1)$$

The brackets comprise longitudinal (z), azimuthal (Θ) and radial (r) components. The indices M and F indicate matrix and fibers, respectively.

ε_M^f can be increased by

- large fracture strain of the matrix
- adapting the thermal contractions of fiber and matrix and
- creep relaxation down to the lowest possible temperatures.

The latter solution is helpful to some degree only since a certain creep stability is necessary at RT. However, in many low temperature applications the structural elements made from composites mainly have to carry magnetic forces. In the case of superconducting magnets they only act at very low temperatures where creep is negligible. At RT those structures have to carry only their intrinsic weight.

FRACTURE STRAIN OF THE MATRIX ε_{UTM}

Most polymers show a low fracture strain at cryogenic temperatures. Their stress-strain relation is linear till fracture occurs. The static fracture strain ε_{UTM} at 4.2 K for most polymers does not exceed 2%. A list of data is given in Table I.

Several investigations have shown that cross-linked polymers such as

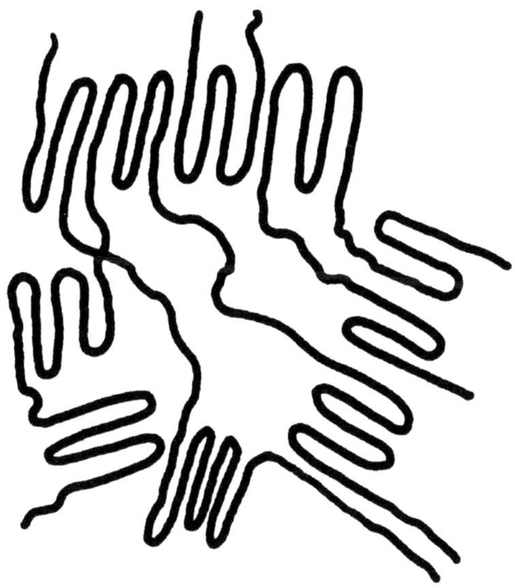

Fig. 2. Tie-molecules between crystalites of HDPE

epoxy resins are bound to cryogenic brittleness. The types of epoxy resins commonly used for prepreg technique exhibit a fracture strain below 1.6% at 4.2 K. There are some linear thermoplastic polymers, namely PC, PSU, PES and HDPE, which show some low temperature ductility even at 4.2 K. Their stress-strain diagrams are non-linear, as shown in Fig. 1 for PC and HDPE.

The question arises as to which fracture mechanisms are dominant in those polymers:

a) Low temperature relaxations
All low temperature relaxations ($T < 40$ K) so far investigated have their origin in the mobility of side- or end-groups which do not influence the stress-strain behavior nor the creep characteristic within finite times.

b) Collective motion of tie-molecules
The amorphous tie-molecules in partial crystalline polymers (e.g. HDPE) exhibit a rather uniform load transmission into the crystallites. An illustration is given in fig. 2. Thus, under load, stress concentrations on a single molecule may be less than in the case of randomly entangled molecules of amorphous polymers. It should be emphasized that this statement is not sufficient for ductility, since not all crystalline polymers show this behavior. Even HDPE with a very high molecular weigth $> 5 \cdot 10^6$ is brittle at 4.2 K. Best results have been found for molecular weights of $\approx 5\ 10^5$ g/mole.

c) Craze opening
Several thermoplastic polymers tend to a craze formation at tips of microcracks during fabrication or external load. As pointed out in the paper by W. Döll[4] the crazes consist of strong fibrils oriented in the direction of local load. Crazes at crack tips are thought to increase the fracture toughness, the craze formation is a rather energy consuming process (absorption of fracture energy).

According to interference measurements the craze opening can be rather large (up to 100%). The craze opening is mainly an elastic reversible process, whereas load induced craze formation is inelastic. It seems that both processes take part in the low temperature ductility of several thermoplastic polymers. For polycarbonate (PC): $\varepsilon_{UTM}(4.2K) \approx 3.5\%$. PSU and PES show somewhat lower values (see Table I). However, the question remains to be clarified as to why only some of the "crazing" polymers show a low temperature ductility. PVC or PMMA, e.g., tend to craze formation, but they show a linear stress strain behavior and a fracture strain $\varepsilon_{UTM} \approx 2\%$ at 4.2 K like for most polymers.

THERMAL RESIDUAL STRAIN ε_M^R

For unidirectional* fiber composites it holds for the matrix strain at the fiber surface[5]:

$$\varepsilon_{zM}^R = \int_{T_1}^{T_2} \frac{1}{E_M(T)} (\sigma'_{zM} - \mu_M(\sigma'_{\Theta M} + \sigma'_{rM}))\, dT \qquad (2)$$

analogue for $\varepsilon_{\Theta M}$, ε_{rM}

$\sigma' = d\sigma/dT$ are the thermal residual stresses per temperature difference.

$$\sigma'_{zM} = E_M(T)\ (\alpha_M - \alpha_{zF}) - 2\ \mu_M \cdot P_s \cdot \frac{\varphi}{1-1.1\varphi} \qquad (3)$$

$$\sigma'_{\Theta M} = -\ P_s \cdot \frac{1+1.1\varphi}{1-1.1\varphi} \qquad (4)$$

$$\sigma'_{rM} = +\ P_s \qquad (5)$$

$$\text{shrinkage compression: } P_s \approx E_M(T)\ (\alpha_M - \alpha_{rF}) \cdot \frac{(1 + \mu_M)}{\frac{1+1.1\varphi}{1-1.1\varphi} + \mu_M} \qquad (6)$$

E: young's modulus; poissons's ratio μ

φ: fiber content by volume

Equation (6) holds only for medium values of φ and $E_{rF} >> E_M$.
α is the thermal coefficient of expansion.

The integral thermal contraction $\Delta L/L = \int \alpha dT$ is plotted for two polymers in Fig. 3.

Fiber glass is isotopic in its mechanical and thermal properties, whereas carbon-and Kevlar fibers are highly anisotropic.

For the components of the Young's modulus and the thermal contraction it holds:

Fiber glass: $E_{zF} = E_{\Theta F} = E_{rF} >> E_M$ and $\alpha_{zF} = \alpha_{\Theta F} = \alpha_{rF} << \alpha_M$

Carbon fibers: $E_{zF} >> E_M; E_{\Theta F} \approx E_{rF} \gtrsim E_M$ and $\alpha_{zF} << \alpha_M;\quad \alpha_{\Theta F} \approx \alpha_{rF} \lesssim \alpha_M$

---

* The residual and thermal residual stresses, especially shear stresses in angle ply composites are beyond the scope of this paper. The thermal residual stresses in directions of maximal load can be minimized by proper fiber arrangements.

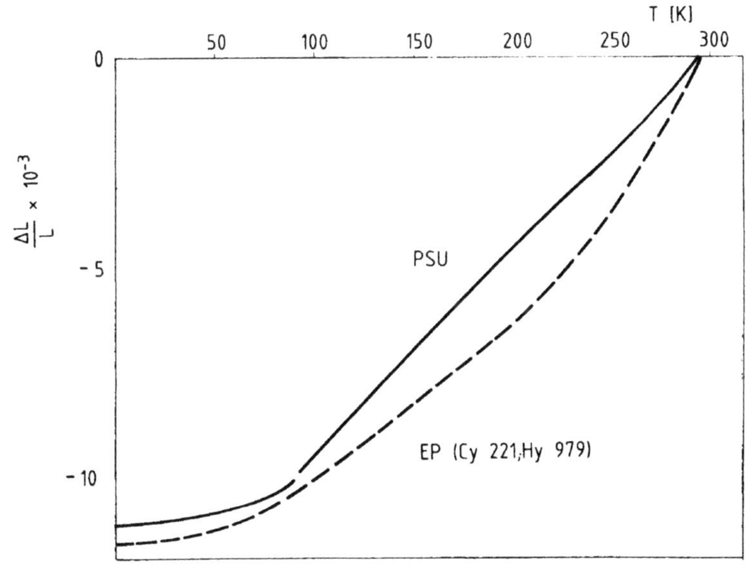

Fig. 3 Integral thermal expansion versus temperature of polysulfone and an epoxy resin (Cy 221, Hy 979)

Kevlar fibers: $E_{zF} >> E_M; E_{\Theta F} \approx E_{rF} \approx E_M$ and $\alpha_{zF} << \alpha_M$; $\alpha_{\Theta F} \approx \alpha_{rF} \gtrsim \alpha_M$

<u>For carbon- and Kevlar fibers</u> the rough approximation is made:

$\alpha_{zF} \approx 0$ and $\alpha_{rf} \approx \alpha_M$. Then the shrinkage compression P_s vanishes and it holds for the remaining z-component of ε_M^R:

$$\varepsilon_{zM}^R \approx \int_{T_1}^{T_2} \alpha_M \, dT = \left(\frac{\Delta L}{L}\right)_M (T_2 \rightarrow T_1) \qquad (7)$$

Assuming a temperature difference $T_1 = 293$ K and $T_2 = 4.2$ K one gets by means of Table I:

Matrix	Free strain ε_{zM}^f	Free strain ε_{zM}^f with creep
Prepreg epoxy resin	< 0.5	< 0.5
Laminating " "	1.0	1.1
PC	2.1 *	2.0
HDPE	$\approx$ 2.3	$\approx$ 2.5

These values are compared with the fracture strain ε_c of unidirectional carbon- and Kevlar fiber composites as summarized in Table II. One finds that for high tensile carbon fiber composites the fracture strain is about 1.6% and 1.5% for Kevlar fibers. This means that an epoxy matrix fractures much earlier than the fibers. It should be borne in mind that new developments in the carbon fiber production will increase the fracture strain up to 1.8%. Due to a higher ε_M^f, composites with a PC matrix are on the safe side.

For dynamic loads the situation is even worse, since the fatigue endurance limit of polymers at cryogenic temperatures is - as a rule of thumb - about 40% that of the static fracture strength, whereas for carbon- fiber composites it amounts 85% (tensile threshold load).

Fig.4a- Fractograph of (0°,±45°90°) angle-ply composite with M 40A carbon fibres at 77 K. b-Fractograph of UD carbon fibre composites at 77 K (M 40A-carbon fibres)

For isotropic fiber glass $\alpha_{zF} = \alpha_{\Theta F} \approx \alpha_{rF} << \alpha_M$ and P_s is non negligible and a multiaxial stress situation takes place. But it can be shown by eqs. 1 to 6 that for $E_{zF} = E_{\Theta F} = E_{rF} >> E_M$ the longitudinal component is nearly not influenced by other strain components. The thermal residual strain on the matrix between T_2 = 293 K and T_1 = 4.2 K is calculated for composites with a relative fiber content of φ = 0.6 by volume by eqn's 1) to 6). The integral thermal contraction of fiber glass is $\Delta L/L_F$(293K-4.2K)$\approx$0.18%.

Matrix	ε^R_{zM}	ε^f_{zM} without creep
Prepreg epoxy resin	1.0%	0.6%
Laminating " "	1.0%	1.1
PC	1.2	2.3*
HDPE	1.6	1.4-2.4

* The free strains of polycarbonate and polysulfone are nearly the same. Polycarbonate has a larger ultimate tensile strain but a larger thermal expansion than polysulfone.

Table I. Mechanical and Thermal Data of Selected Polymers

	FRACTURE STRAIN %		FRACTURE STRENGHT MPA		YOUNG'S MODULUS GPA		POISSON'S RATIO	$\Delta L/L$
	4,2 K	77 K	4,2 K	77 K	4,2 K	77 K		293K-4,2K
PREPREG EPOXY RESIN	<1,6	<1,6	—	—	—	—	—	≈-1,1%
LAMINATING RESIN	2,1	3,2	175	—	8,2	7,6	—	≈-1,2%
POLYCARBONATE PC	3,5	4-6	170	150	6,5	5,2	0,39	≈-1,4%
POLYSULFONE PSU	3,2	4,2	164	120	5,6	4,8	0,413	≈-1,1%
POLYETHERSULFONE PES	2,8	5,0	144	160	—	5,9	—	—
POLYETHYLENE HDPE	3-4	—	190	—	9,8	—	—	≈-1,8%

Table II. COMPOSITES

	ε_c (4,2K)
FIBER GLASS COMP.	1,8-2,5 %
CARBON.TH COMP.	1,6 (1,8) %
KEVLAR COMPOSITES	1,3-1,5 %

As seen from Table II the fracture strain of unidirectional fiber glass composites is about 1.8 to 2.5%. Thus, an epoxy resin matrix fractures much earlier than the fibers. For thermoplastic matrices like PC and PSU the situation is balanced. The situation is not stressed under dynamic load since the fatigue endurance limit of fiber glass composites is lower than that of a polymer.

FAILURE MODES IN ANGLE-PLY COMPOSITES

Among various failure modes, such as delamination, one failure mode will be considered which is typical of brittle polymeric matrices. When angle-ply composites (0°, $\pm 45^{o}$, 90°) are loaded - say in the 0° direction -crack initiation and propagation are favored within 90° and $\pm 45^{o}$ layers. When cracks reach a load bearing layer (0°) they cut fibers as soon as the stress concentration exceeds their strength. This is true for transverse weak fibers such as carbon - or Kevlar fibers. Fiber glass is rather tough and acts as a crack stopper. In Fig. 4a a fractograph of such a carbon fiber composite shows a plain and smooth fracture area produced at 90° and a stepwise fracture area at 45° in the load bearing 0° layer. By contrast, the rough fracture area with fiber pullout of UD-composites is shown in Fig. 4b. Although the matrix has negligible load bearing functions, it can reduce the strength of a carbon fiber composite by 30% and more. This fracture mode could be avoided by a matrix which is ductile at low temperatures.

CONCLUSION AND OUTLOOK

In cryogenic applications the brittleness of epoxy resins as the matrix reduces the mechanical strength, espcecially under multiaxial load conditions. The study of degradations of composite properties under fatigue cycling is a subject of further investigations. For future developments a substitution of brittle epoxy resins by special low temperature ductile thermoplastics will help to reduce thermal residual stresses and stress concentrations within fiber composites. There are other advantages of thermoplastic matrices: they are hot workable and they can be "welded" by solvent agents or by heat. For epoxy resins after curing an adhesion (by secondary bonding) is possible only. Current problems like fabrication (prepreg technique) or the fiber-matrix adhesion which controls the interlaminar shear strength will be considered in the next two papers.

REFERENCES

1. Hartwig, G.: from Proceedings of the International Cryogenic Materials Conference, Kobe, Butterworth (1982) pp. 495-502.
2. Hartwig, G. and S. Knaak: Cryogenics, Nov. 1984, Butterworth 1984, pp. 639-647.
3. Hartwig, G., Jäger, H., and Knaak, S (in this issue)
   E. Fitzer, G. Hannes and H. Jäger (in this issue).
4. Döll,W.: Advances in Polymer Science 52/53, Springer-Verlag Berlin, Heidelberg, (1983), pp. 106-165.
5. Schneider, W.: Kunststoffe, Vol. 61 (1971), pp. 273-277.
6. Hartwig, G.: Advances of Cryogenic Engineering (Materials), Vol. 28, Plenum Press, London and New York (1982), pp. 179-189.

FABRICATION OF CARBON FIBRE REINFORCED

THERMOPLASTICS POLYCARBONATE AND POLYSULFONE

E. Fitzer, G. Hannes, and H. Jäger

Institut für Chemische Technik
Universität Karlsruhe
Kaiserstr. 12, D-7500 Karlsruhe

In recent years the market for advanced composite preparation was oversupplied by thermoplastics as a new group of matrix systems.[1-4] Within these polymers polycarbonate PC and polysulfone PSU seem to be most suitable for carbon fibre reinforced thermoplastics at cryogenic temperature application because of their highest tensile strains up to 4.2 K.[5] In this temperature range sufficient adhesion between carbon fibre and thermoplastic matrix is only achieved if the contamination of the active carbon fibre surface by water adsorption is hindered. A preparation method is presented that offers the possibility to form composites with thermoplastics PC and PSU by prepreg technique. Controlling quality parameters like pressure and press temperature are discussed with respect to the properties of the carbon fibre reinforced thermoplastic composites.

In advanced composite preparations with reactive matrix systems like epoxies, no infiltration and hardening problems occur because of the liquid state of those resins. In Fig. 1 (left hand side) the behaviour of a commercial epoxy mixture with amine hardener in dependance of time is shown. One can clearly recognize the decrease of viscosity to a minimum after 30 min. With increasing temperature cross-linking reaction and the formation of networks lead to an increase in viscosity.[6] But in any case viscosity of mixtures in liquid state is not so high as that of thermoplastics PC and PSU (right hand side). Only at temperatures above 220 °C in case of PC and more than 300 °C for PSU, viscosity of PC and PSU reduces so that processing seems possible at higher pressures.

Beside of this problem with matrix work up, the contamination of the polar C-fibre surface by water adsorption is the next problem, overcoming producers of composites. In Fig. 2 the formation at first of a monolayer of water before multilayer formation till equilibrium is shown in a simple model. This adsorption of water on the carbon fibre surface can be demonstrated by weight increase of a well-dried C-fibre in a definite climate. For this investigation a high tensile C-fibre without surface finish (Hercules As4 - 12K) was vacuum-dried (1 Pa) for 24 h at 150 °C. After cooling down weight increase at a relative humidity of 63.5 % and 26.5 °C was measured during adsorption phase. We can recognize a very strong nearly linear weight increase within seconds up to about 7 minutes. Therefore, in order to produce composites with good adhesion at the fibre matrix interface, the water contamination should be lowered by heating up the C-fibres before processing.

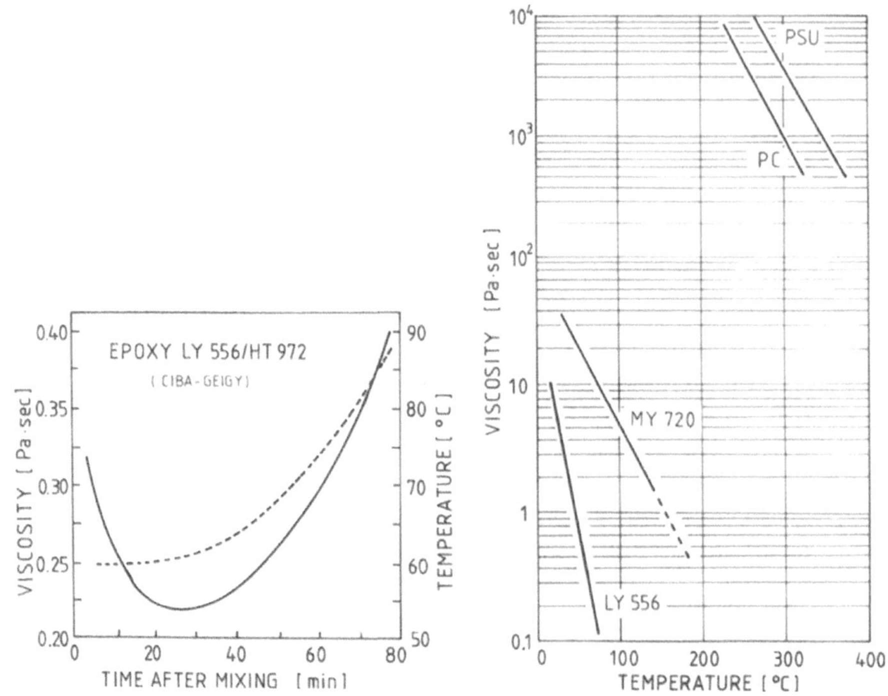

Fig. 1. Viscosity and temperature behaviour of an epoxy system after mixing (left hand side) and comparison of the viscosity of thermoplastics PC and PSU with commercially available epoxies.

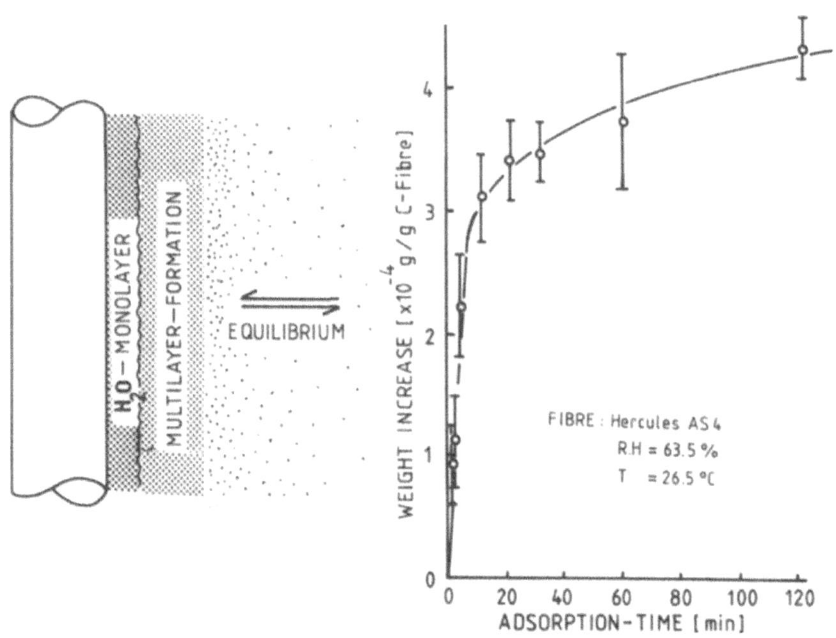

Fig. 2. Adsorption of water on the C-fibre surface, demonstrated by measuring the weight increase with adsorption time.

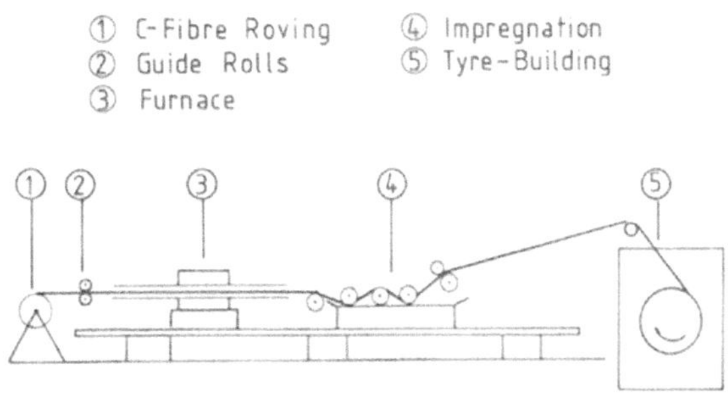

Fig. 3. Plan of a continuous impregnation plant for impregnation of C-fibres with thermoplastics PC and PSU.

TEM OF HERCULES AS4, IMPREGNATED WITH PC

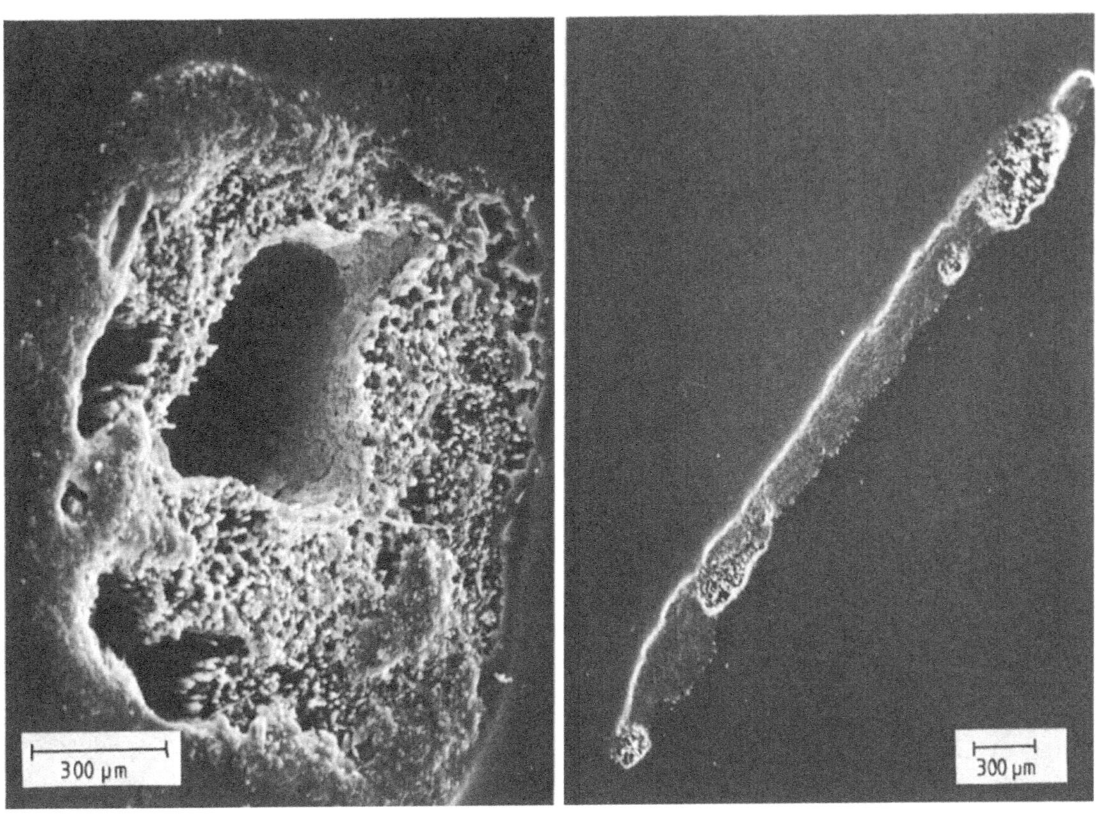

Fig. 4. Impregnated prepregs with correct stretch (right hand side) and too small stretch (left hand side).

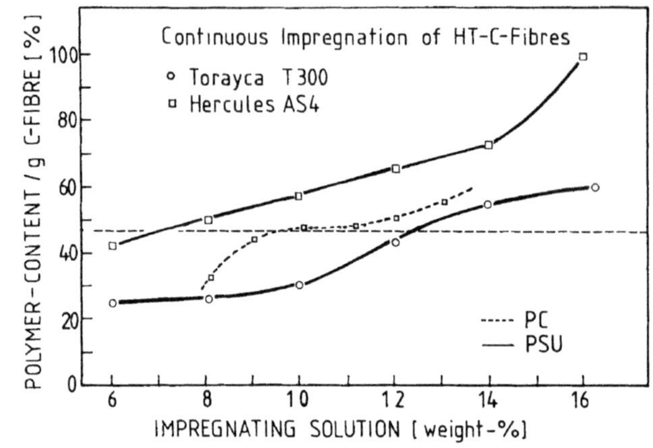

Fig. 5. Influence of various commercially available fibre-types on polymer content after impregnation with solutions of different contents of PC and PSU (in weight %).

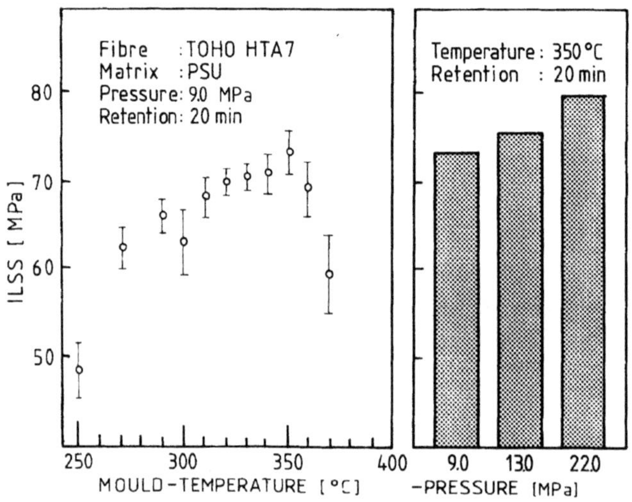

Fig. 6. Studies on ILSS-dependance of composites by different mould temperatures and mould pressures (only PSU).

Furthermore good infiltration of the solid PC and PSU into the fibre bundle can only be achieved by working with their solutions. These preconditions are sufficient to build a continuous impregnation plant as shown in Fig. 3. Here C-fibre is heated up in a furnace (250 °C, inert atmosphere) to remove rests of adsorbed water before impregnation. In prepreg formation the impregnation step is most important because of the infiltration of the bundle by impregnating solution. If roving stretch is too small, the fibre impregnation leads to tubes (see Fig. 4, left hand side). In case of enough stretch, well plane impregnated prepregs are formed (see Fig. 4, right hand side). These plane prepregs guarantee best fibre orientations during pressing.

The dominant fibre surface effect on polymer content during impregnation is demonstrated in Fig. 5. The dashed line indicates the optimum polymer content of the dried prepregs for composite formation. As can clearly be recognized each fibre type needs a special thermoplast concentration in $CHCl_3$ (CH_2Cl_2) - solution to achieve sufficient polymer excess for the following pressing procedure. For the Hercules AS4 C-fibre (12K) about 6 % PSU - solution leads to good prepregs whereas with Torrayca T300 (6K) more than 12 % of PSU are needed. With AS4 fibre in

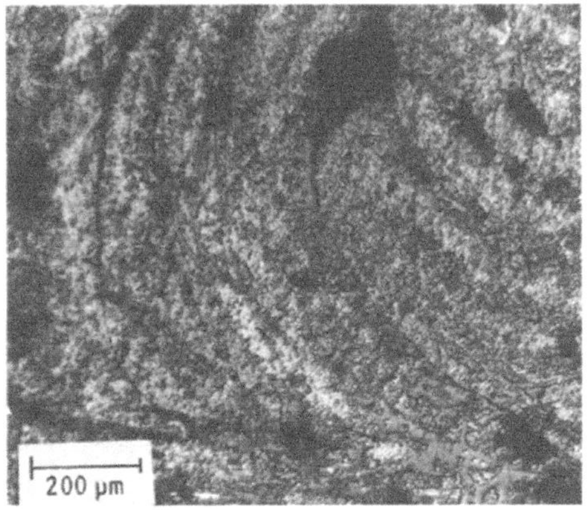

T = 250 °C, p = 9.0 MPa

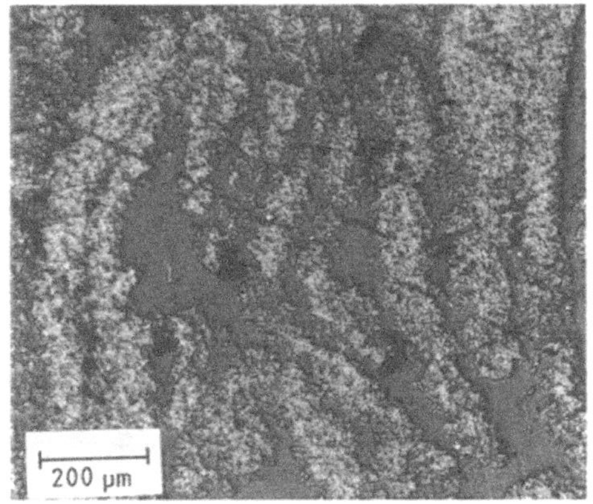

T = 290 °C, p = 9.0 MPa

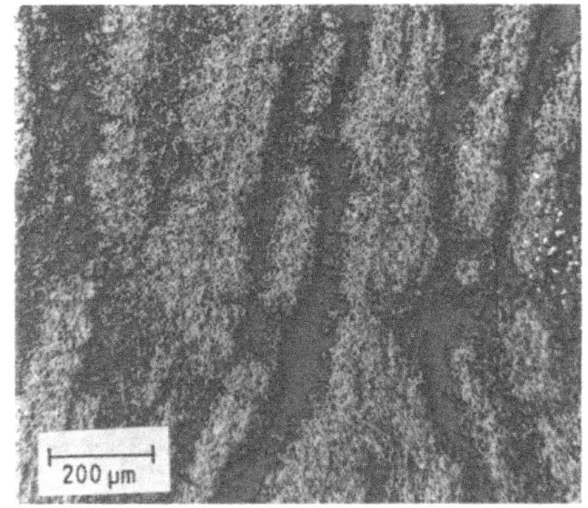

T = 350 °C, p = 9.0 MPa

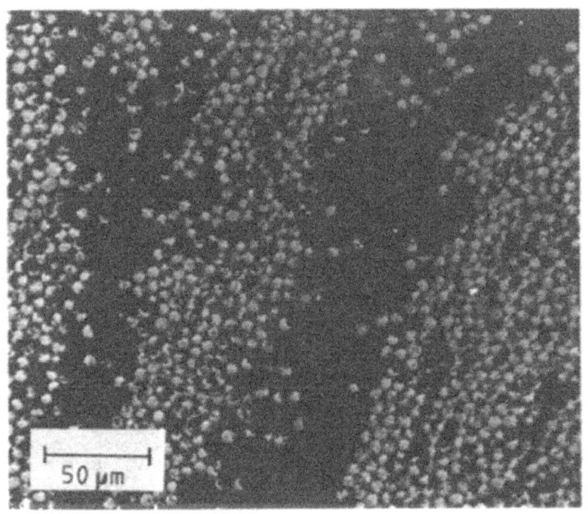

T = 350 °C, p = 9.0 MPa

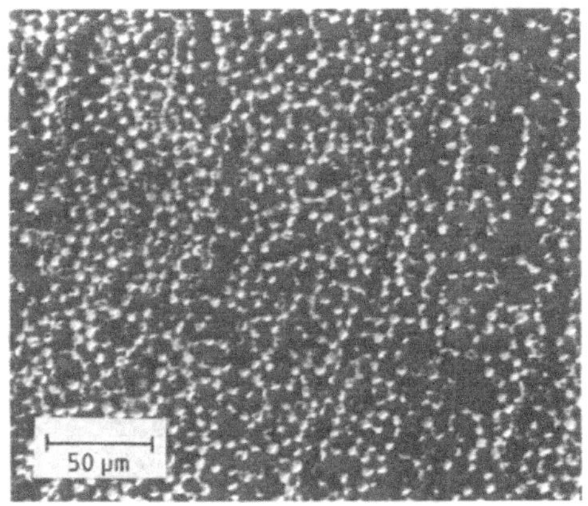

T = 350 °C, p = 13.0 MPa

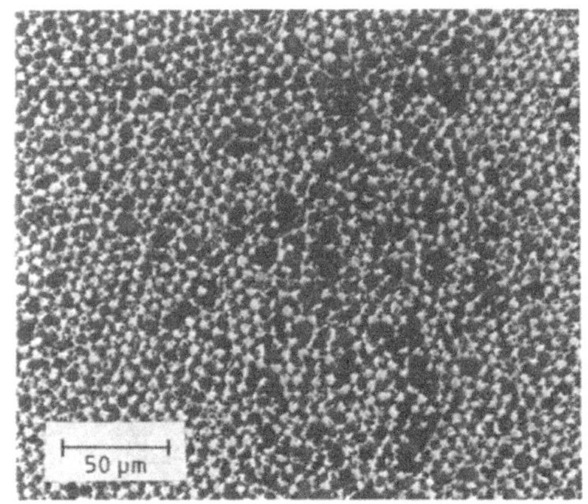

T = 350 °C, p = 22.0 MPa

Fig. 7. TEM of composites during different optimization steps (as described before).

PC solution best results are obtained with impregnating solution of about 10 %. This indicates that beside of the fibre type the type of solution (PC or PSU) estimates the prepreg quality.

A further controlling parameter in prepreg technique is the drying cycle of the impregnated rovings. Optimum composite behaviour is obtained by the following mode: Drying for 12 hours at RT in air, for 20 hours at increasing temperature up to 120 $^{o}C_{PC}$ (165 $^{o}C_{PSU}$) under reduced pressure (up to 1 Pa).

Regarding carefully all these described parameters, best prepreg qualities are obtained.

To form now UD-model composites by prepreg technique knowledge about optimum mould-temperature and mould-pressure is necessary. During these investigations a temperature distribution within the mould (length of 18 cm) was estimated. Because of this effect, only small middle parts (9.5 cm length) of the produced composites could be used for further studies. For the matrix PSU as an example (see Fig. 6) the way of optimization of composite technology is described. At one pressure, a variation of mould-temperatures is made with the aim to achieve best mechanical composite properties (retention time in all these studies was 20 minutes). Maximum ILSS of 73 MPa was obtained at 350 ^{o}C mould temperature and a constant pressure of 9.0 MPa. At this optimized temperature, pressure was varied and resulting ILSS was measured, too. Best conditions to form composites with PSU as matrix were found to be 350 ^{o}C and 22.0 MPa pressure. (ILSS of about 80 MPa). The same pressure but 245 ^{o}C gives best conditions for PC as matrix (estimated in the same manner).

The effect of optimization process on fibre distribution in composite is shown in Fig. 7. Only optimized conditions give a homogeneous C-fibre distribution. Therefore in case of homogeneous infiltration, best mechanical composite properties are achieved.

CONCLUSIONS

Contamination of the C-fibre surface by adsorption of water during processing must be hindered by a pre-heating of the fibres.

To achieve best matrix properties drying cycle should be optimized so that no decrease in matrix strength by lower glass transition point occurs.

Only at optimized mould-temperature and -pressure maximum mechanical composite properties can be achieved.

REFERENCES

1. M. Buhl, U. Neumann,"Carbonfaserverstärkte Thermoplaste,"Plastverarbeiter 31,9 (1980), pp. 524-526.
2. I. Brewster, J.B. Cattanach,"Engineering with long fibre thermoplastic composites," SAMPE, European Meeting 1983, London, paper 3.
3. A. P. Cluley,"Carbon fibre reinforced thermoplastics high performance engineering materials,"SAMPE, European Meeting 1983, London, paper 5.
4. K. J. Humphris,"Thermoplastic composites: a growing family of new materials,"Shell Polymers, vol. 7, no. 1 (1983).
5. G. Hartwig, H. Jäger, S. Knaak, "Interlaminar shear strength of carbon fibre reinforced thermoplastics polycarbonate and polysulfone", compare this issue.
6. M. Fischer, F. Lohse, R. Schmid, "Struktureller Aufbau und physikalisches Verhalten vernetzter Epoxidharze," Makromol.Chem. 81,(1980), pp. 1251-1287.

INTERLAMINAR SHEAR STRENGTH OF CARBON-FIBRE REINFORCED THERMOPLASTICS POLYCARBONATE AND POLYSULFONE

G. Hartwig, H. Jäger*, and S. Knaak

Kernforschungszentrum Karlsruhe, Institut für Reaktor-bauelemente
*Universität Karlsruhe, Institut für Chemische Technik
Karlsruhe, Federal Republic of Germany

ABSTRACT

The interlaminar shear strength (ILSS) of carbon-fibre reinforced polymers (CFRP) at room temperature (RT) depends on:

a) properties of the reinforcing fibre
b) properties of the pure matrix system
c) interaction between fibre surface and matrix, caused by active surface groups

At cryogenic temperatures two effects are superposed:

a) increasing matrix strength
b) "shrink-on" of the matrix on the fibre because of the strong matrix contraction by cooling

Both effects can cause improved mechanical composite properties. Negative side-effect of this shrinkage is the built-up of enormous thermal residual stresses at the fibre-matrix interface and therfore a more brittle fracture behaviour. Matrix systems having some low temperature ductility and high cryogenic fracture strain are necessary to overcome these problems.

Thermoplastic materials like polycarbonate (PC), polysulfone (PSU) and polyethersulfone (PES) seem to be promising candidate matrix materials.

The ILSS values, measured by short beam test on PC and PSU reinforced with HM- and HT-fibres at RT, 77K and 4.2 K are presented and discussed. A comparison is given of the different fracture behaviour of carbon fibre reinforced epoxies, polycarbonate and polysulfone.

INTRODUCTION

The composite properties of carbon fibre reinforced polymers at room temperature depend on the fibre properties, the properties of the used matrix system and the so-called fibre-matrix interaction.[1]

That means the influence of the reactive side-groups of the matrix and the active surface groups of the carbon fibre on the adhesion and infiltration in composite. Each of these described parameters contribute to the CFRP and all together guarant best composite behaviour.

In Table I we can compare the carbon fibre properties of the fibres used in our experiments. As high tensile fibres we selected those of TORAYCA (T300) and HERCULES (AS4), as a high modulus type TORAYCA M40. All these types were PAN - based. T300 with a medium fibre strength of 3100 MPa and a strain to failure of 1.5%, on the other side the high strain AS4 - fibre with strength of 3800 MPa and strain of about 1.7% are compared with the M40, a high temperature treated C-fibre with a Young's modulus of about 400 GPa (HT-types result only in 230 GPa). All these used fibre-types are well surface treated, that means the fibre surface had been activated by a special oxidation treatment and during this oxidation process surface oxides were formed.

Fig. 1 shows us the different types of surface oxides built up during this oxidation step.[2] The evidence of these groups can be shown by desorption method[3] (heating up to more than 1000 °C in a short time and analysis of the splitted groups by IR - method and weight loss) and by chemical method[4,5,6] (reaction with alkaline and other specific reagents). The last way offers the possibility to explain the chemical structure of the surface groups. We can detect aldehydes, carboxyles, hydroxyles as acidic groups and carbonyles/quinones as basic ones.[7] Most of these groups are able to react chemically, even at RT with the functional groups of the epoxy matrix normally used in CFRP's as thermosetting matrix system. Some of these compounds are unstable under hydrolyse conditions (amines used mostly as hardener) other ones react only at elevated temperatures (more than 100 °C) with epoxies.[8]

The influence of these surface groups especially on the ILSS of composites has been discussed in previous works by Fitzer et al. [9,10]
As can be seen in the left part of Fig. 2 ILSS increase with increasing amount of surface oxides detected by desorption. This is also demonstrated in the upper diagram (right hand side) by measuring the ILSS of untreated and surface treated HM- as well as HT-type C-fibres. Untreated HM-C-fibres result in ILSS of about 20 MPa; in case of optimum surface treated HM-fibres ILSS values of 70 - 80 MPa were achieved. Compared to untreated HM-C-fibres high tensile fibres untreated levels at values of about 50 MPa and surface treatment increase the ILSS only up to 80 MPa. In all these investigations bifunctional epoxy LY556 (CIBA-GEIGY) with amine hardener DDM was applied. The contribution of the surface oxides formed during oxidation process to these ILSS values could be explained by blocking experiments, that means desactivation of these surface oxides chemically without changing roughness of the surface area (measured by BET). Using pure CH_2N_2, ILSS of treated fibres (HM) decrease to the level

Table I.

CARBON-FIBRE PROPERTIES

Fibre	Type	Tensile Strength [MPa]	Strain to failure [%]	Young's modulus [GPa]
Torayca T 300	HT	3100	1,30	230
Hercules AS4	HT	3800	1,69	230
Torayca M 40	HM	2500	0,60	400

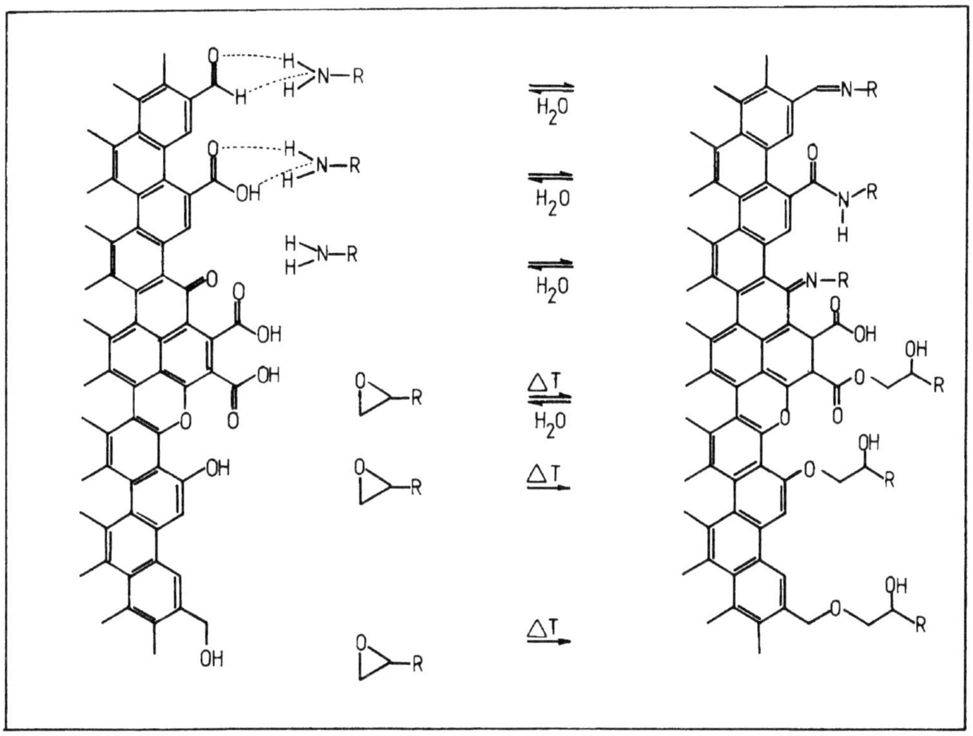

Fig. 1 Types of surface oxides formed during oxidation process and reaction possibilities with epoxy and amine.

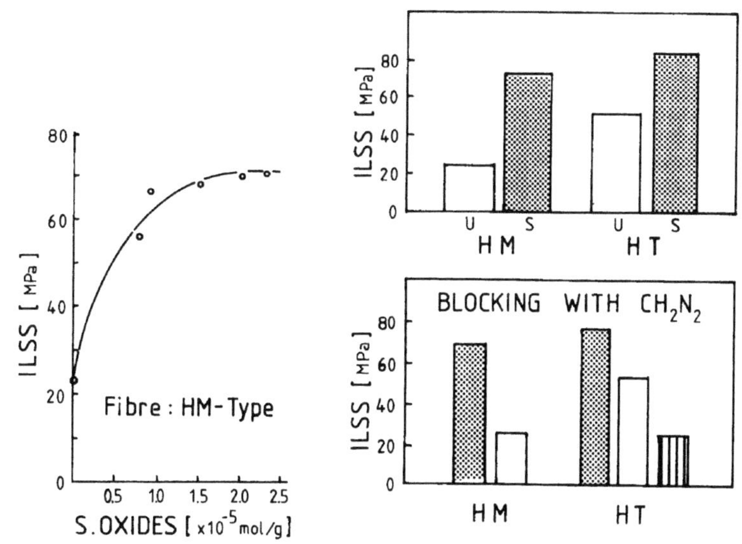

Fig. 2 Influence of surface oxides on the ILSS of CFR-epoxy.

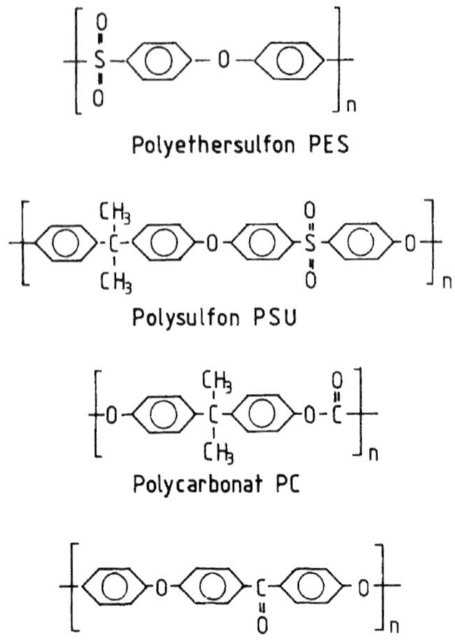

Fig. 3 Structure of thermoplastics.

of the untreated ones whereas only with contaminated DIAZOMETHANE all reactive HT-C-fibre surface groups could be desactivated.

By these blocking experiments in case of using reactive matrix systems like epoxies with chemical reaction possibilities between matrix functionalities and surface oxides, the manner of adhesion could clearly be demonstrated.[6]

For thermoplastics such as POLYETHERSULFONE PES, POLYSULFONE PSU and POLYCARBONATE PC (see fig. 3) no direct chemical reaction possibility is given. The sole possibility is the reaction by 'VAN DER WAALS' bonds with the polar groups of the thermoplastic matrix systems.

The adhesion in CFRP's can quantitatively be measured by shear test of the composites (see Fig. 4). A simple test method here is the three point short-beam test according ASTM (left hand side).[11] However in composites reinforced with surface treated C-fibres often no pure delamination is observed. But this is a precondition for an exact determination of the real ILSS-values. We used this test method because of simple application in a very complicated test apparatus at low temperatures. An improved testing procedure is shown at the right hand side of Fig. 4. It is a tensile torsional test developed for tubes[12]. By this method exact ILSS values can be determined. But the preparation of the filament wound test tubes is very time consuming and the achieved values don't differ strongly from results measured by the easy three point short beam test.

To demonstrate the adhesion in the thermoplastic systems PC and PSU at RT composites were tested with HM- and HT-type C-fibres TORAYCA and HERCULES (mechanical properties see Table II). In both thermoplastic systems with HM-C-fibres ILSS values of about 50 MPa were achieved whereas in case of reinforcement with HT-fibres, ILSS levels at about 70 - 80 MPa.

At cryogenic temperatures the adhesion in composites is estimated by two further parameters: the higher matrix strength and the possible shrinkage of the matrix on the fibre.[13] In radial direction, as shown

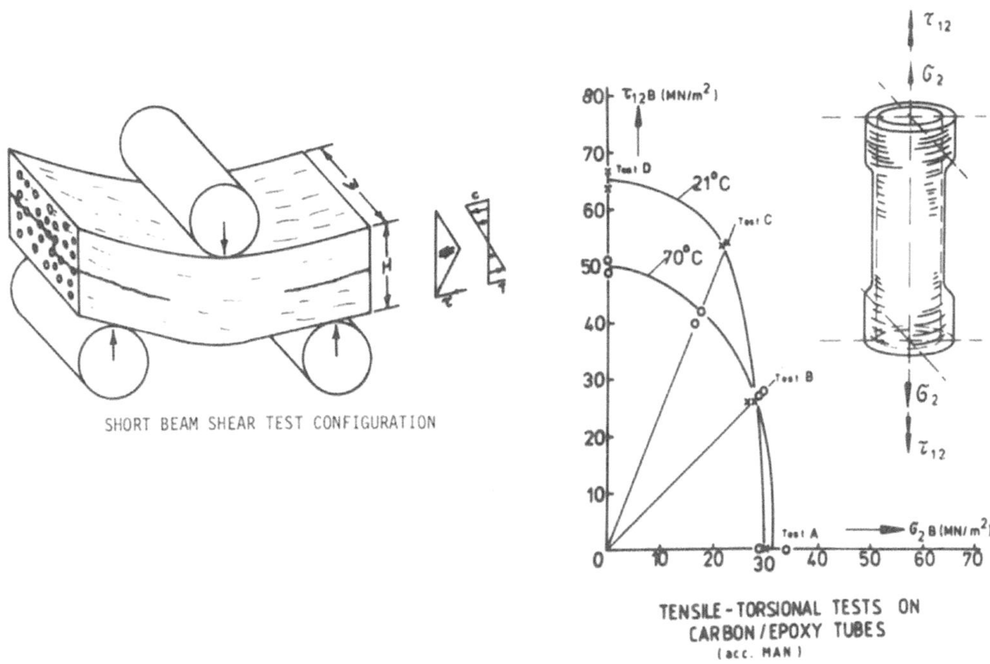

Fig. 4 Test methods to estimate the ILSS of composites.

by G. HARTWIG (paper in this issue), nearly no stress-built-up occurs in composites with carbon- or Kevlar-fibres. In fibre direction (longitudinal) there is a negative side effect by the formation of enormous thermal residual stresses at the fibre-matrix interface and therefore a more brittle fracture behaviour.

At RT and at longer post-curing times there is a lower stress between fibre and matrix. By cooling up because of the differences in thermal expansion coefficients in fibre direction, matrix contraction is stronger than that of the C-fibres.

This built-up of stresses between fibre and matrix can be reduced by matrix systems having low temperature ductility and high cryogenic fracture strain (see preceding paper).

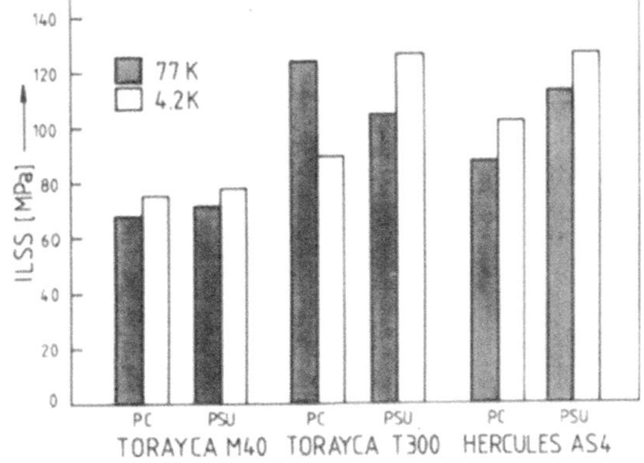

Fig. 5 ILSS of composites with PC and PSU as matrix and HM- (M40) as well as HT-C-fibres (T300), AS4) as reinforcing fibres.

Table II.

MATRIX		MY720/ HT976	LY556/ HT972	CY221/ HY979	PSU	PC
STRAIN TO FAILURE [%]	298 K	1.8	7.0	20	40	60
	77 K		3.0	3.2	6.0	7.0
	4.2 K		2.0	2.1	3.2	3.5

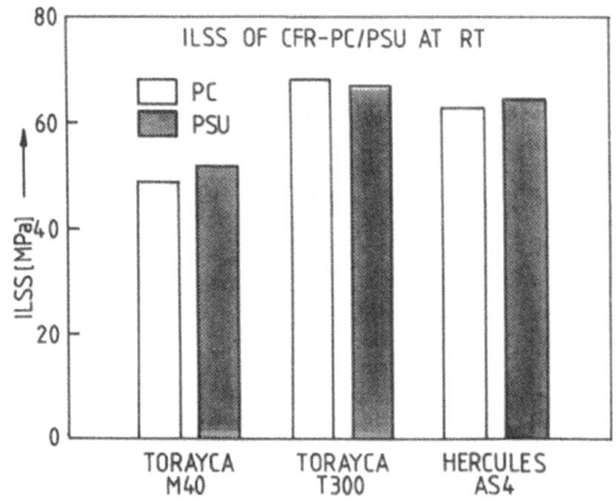

Fig. 6 ILSS of CFR-PC and -PSU at 77K and 4.2K

If we compare the various commercially available epoxy matrix systems with the thermoplastics PC and PSU (see Table II) one can recognise the higher strain to failure of these thermoplastics in all regarded temperature ranges. That means that higher thermal stresses at the interface are tolerated by the thermoplast (see preceding paper). This was the original reason to introduce thermoplastics as matrix materials in composite preparation.

Using epoxies, in most cases ILSS increases at low temperatures. The same effect should be expected with PC and PSU because of the increasing matrix strength.

The ILSS values of composites with HM-C-fibres levels for both matrix systems at 77 K at about 70 MPa. For the HT-fibre T300 in combination with PC highest ILSS of 130 MPa was obtained. With the same fibre and PSU as matrix 110 MPa were achieved. Changing the carbon-fibre with HERCULES AS4 as reinforcing fibre and PC as matrix ILSS of 90 MPa and with PSU of 120 MPa were measured.

In all cases an increase in ILSS at 77 K was achieved. But it seems that special thermoplastic matrix systems are adapted to a single C-fibre-type.

Further increase in ILSS of about 10% by cooling down 4.2K is observed, except with reinforcing fibre TORAYCA T300 (superior ILSS at 77K). Remarkable is the upper level of about 80 MPa for HM-C-fibres as reinforcing fibres. The same value is obtained with nearly all matrix/HM-C-fibre combinations. This effect can be explained with the relatively fibre properties compared to those of HT-C-fibres. Tensile and transverse strength of the high temperature treated HM-C-fibre is lower because of the high degree of orientation of graphitic layers within the carbon-

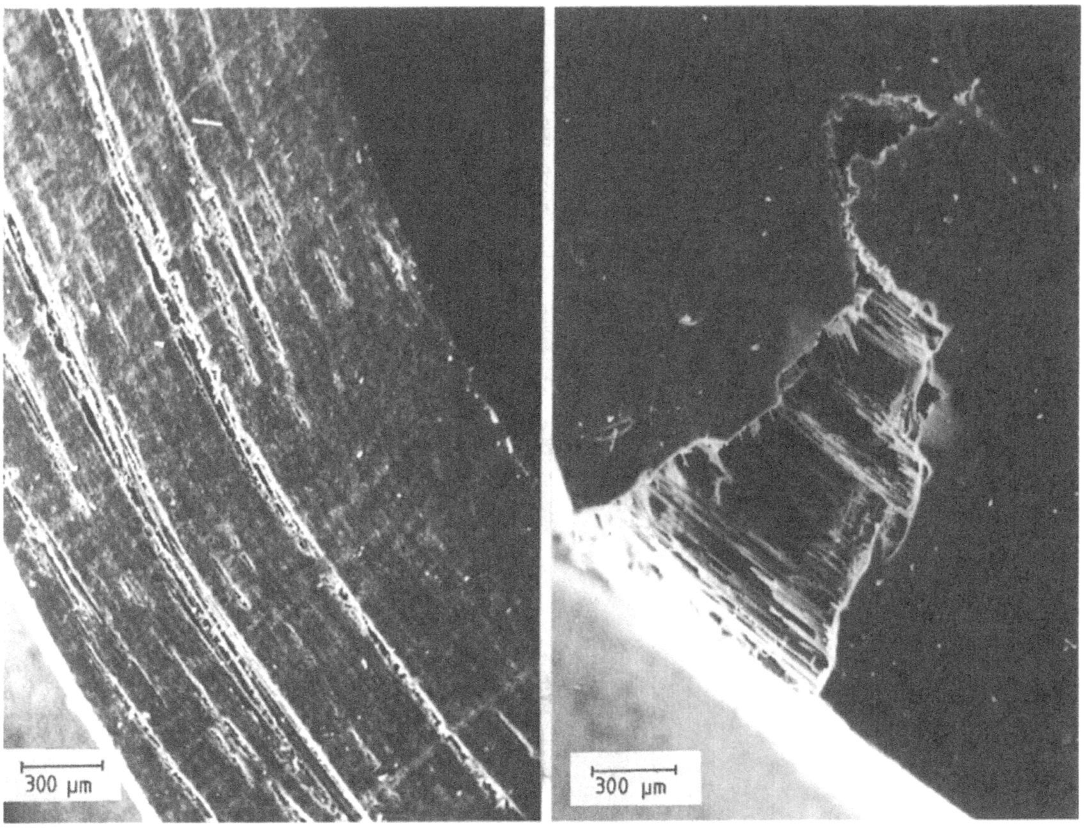

Fig. 7 TEM of ILSS-fracture of CFR-PC: Reinforcing fibre HERCULES AS4 (HT), left hand side, TORAYCA M40 (HM), right hand side.

fibre. The inversion of adhesion in case of T300/PC-combination can be explained by changing of fracture mode. Whilst at 77K only deformation of composite is observed, at 4.2K brittle fracture occurs. All other tested samples reinforced with HT-C-fibres show only deformation whereas with HM-C-fibres (in contrary to epoxy as matrix) fibre pull out can be detected after transverse cracking.

The resistant deformation with all HT-C-fibres, not explained till now, is shown in Fig. 7 (left hand side) in case of using AS4/PC-combination. The right part of the picture demonstrate the fibre pull out of HM-C-fibres. Both specimens were tested at 77K.

For comparison ILSS of composites with HERCULES AS4 as reinforcing fibre and various commercially available epoxies and PC/PSU as matrix are presented (see Fig. 8). As can be seen in all cases values of more than 95 MPa in ILSS at 77K were measured. Visuable is the increase of ILSS with reducing temperature for all matrix variations. In general an increase of about 10% of ILSS by further reduction of temperature down to 4.2K was achieved.

If we compare the bifunctional reinforced epoxy LY556/DDM-matrix system with the above described thermoplastics PC and PSU and as reinforcing fibres HM-type TORAYCA M40 and HT-type TORAYCA T300 (see Fig. 9) the same trend in improving ILSS of composites at low temperatures as with HT-fibre HERCULES AS4 can be observed. The lower ILSS of thermoplastics at RT is equalised at 77K and 4.2K. The low ILSS in general at RT, in reinforcement with HT-C-fibres too, and the enormous increase to values up to 100 MPa in ILSS at 77K and 4.2K (with HM-C-fibres 80 MPa

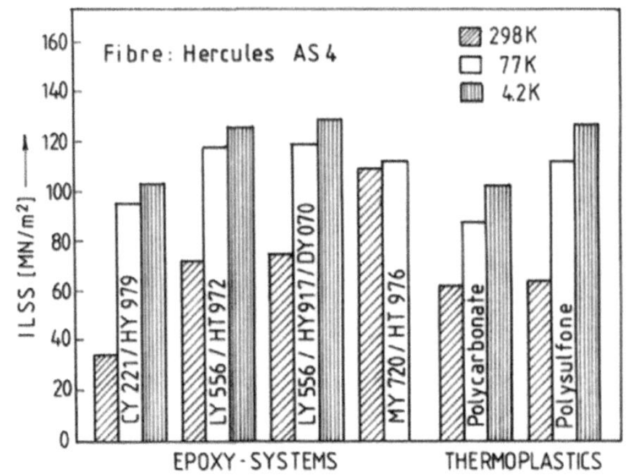

Fig. 8 ILSS of composites with various epoxies and PC as well as PSU as matrix materials (reinforcing fibre: HERCULES AS4) at 298K, 77K and 4.2K.

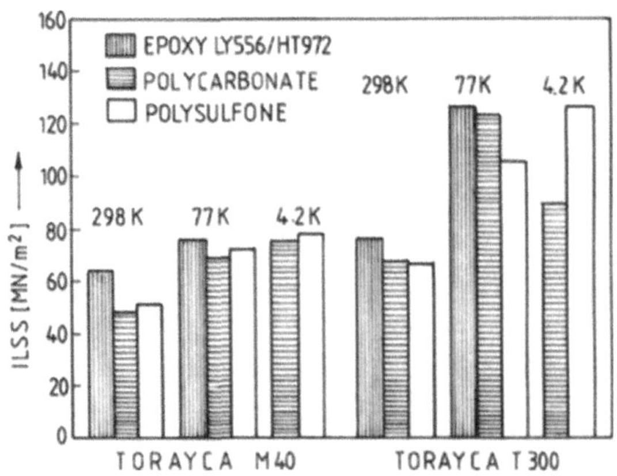

Fig. 9 ILSS of composites with TORAYCA M40 and T300 as reinforcing fibre and epoxy LY556/HT972, PC and PSU as matrix at 298K, 77K and 4.2K.

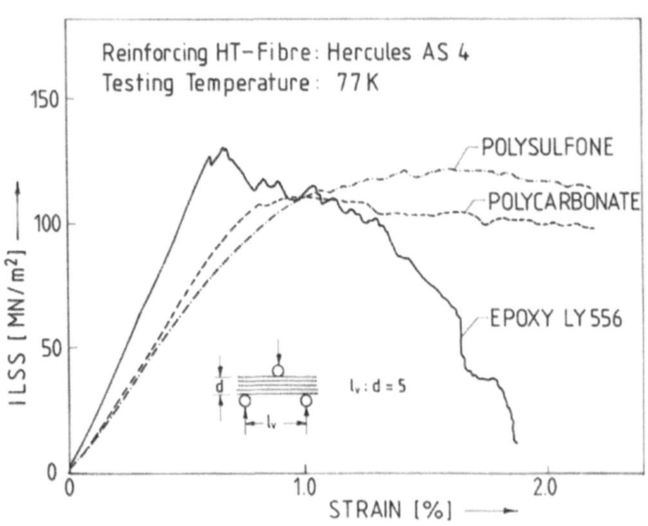

Fig. 10 Stress-strain diagram of HERCULES AS4 reinforced with epoxy LY556, PC and PSU at 77K.

only) confirm the assumption of the dominant effect of the transverse strength of the used fibres.

Summarizing one can say that the same ILSS values of composites with thermoplastics PS and PSU can be obtained as with epoxies. But the enormous advantage of those matrix systems is, as demonstrated by TEM of tested samples (Fig. 7) their high deformability. This behaviour results in composites with remarkable stress-strain diagrams, as shown in Fig. 10. In case of observing brittle fracture with most epoxy-composites at low temperature (except HERCULES AS4 with pure delamination), with thermoplastics as matrix only deformation of the composite and no destruction is observed. This superior fracture behaviour of CFR-PC and PSU indicate the thermoplastics PC and PSU as candidate matrix materials in low temperature application.

CONCLUSION

- The fracture of composites is limited by shear failure of the reinforcing C-fibres (compare HM- and HT-type C-fibres).

- General increase in ILSS of about 10% by temperature reduction from 77K to 4.2K.

- PC and PSU tolerate higher thermal stresses at low temperature.

- Superior fracture behaviour of PC and PSU: only deformation of the composite but no brittle fracture at cryogenic temperatures is observed.

REFERENCES

1. E. Fitzer, R. Weiss, DGM-Tagung Verbundwerkstoffe, Konstanz (1980) p.181-196

2. H.P. Boehme, Angew Chemie 78 (1966), p.717

3. W. Watt, M. Baker, Carbon Conference London (1982), p.359-361

4. H.P. Boehm, Advances in Catalysis 20 (1970), p.179-274

5. D.W. McKee, J.V. Mimeault, Chem. and Phys. of Carbon, P.L. Walker, P. Thrower 8 (1973)p.190

6. E. Fitzer, R. Weiss, Int. Conf. on Interface-Interphase, Liège (1982)

7. G. Trembley, F.J. Vastola, P.L. Walker, Carbon 16 (1978),p.35

8. E. Fitzer, R. Weiss, VDI-Verlag Kunststofftechnik (1981),p.45-67

9. E. Fitzer, K.-H. Geigl, W.Hüttner, R. Weiss, Carbon 18 (1980),p.389-393

10. E. Fitzer, R. Weiss, Carbon Conference San Diego (1983), p.473-475

11. ASTM-D-2344-76

12. Ciba-Geigy, Publ.Nr. 24801/d

13. E.Fitzer, G. Hartwig, R. Weiss, High Temperatures-High Pressures 15 (1983), p.147-150.

ACOUSTIC EMISSION AND INTERLAMINAR SHEAR STRENGTH*

Shigehiro Nishijima and Toichi Okada

Institute of Scientific and Industrial Research
Osaka University
Ibaraki, Osaka, Japan

Shingo Namba

College of Integrated Arts and Sciences
University of Osaka Prefecture
Sakai, Osaka, Japan

ABSTRACT

Acoustic emission from glass cloth reinforced epoxy during interlaminar shear test has been studied in order to clarify the interlaminar properties. At liquid nitrogen temperature the serration was found on load-displacement curves which corresponds with step wise AE increase. At room temperature the macroscopic yielding exist on load-displacement curve which brought the sudden increase of AE energy rate. The detection of interlaminar failure and the estimation of its degree has become possible using AE technique.

INTRODUCTION

Composite materials show complicated and varied fracture process, which results in considerably large scatter in mechanical behavior. The detection and classification method of fracture process should be established in order to estimate the degradation of tolerance or durability. Among fracture processes the interface characteristics is thought to be important because the over-all mechanical properties are strongly affected by the interface behavior that is macroscopic fracture often starts from interface fracture. The initiation of the interface failure should be detected and the degree should be classified in practical application of composite materials.

We have reported that it is possible to classify the fracture processes during the deformation of composite materials using acoustic

---

* Work in part supported by Grant in Aid for Scientific Research No. 59050044 and No. 59750053 Ministry of Education in Japan, and Grant from the Research Project in ISIR.

emission (AE) method.[1,2] In this work the AE technique was applied to the study of the interface characteristics in a composite material.

EXPERIMENTALS

Interlaminar shear tests were performed at room (RT) and liquid nitrogen temperature (LNT) on a commercially available glass reinforced epoxy. The acoustic emissions (AE) were monitored during the tensile test.

Sample

Commercially available glass cloth reinforced epoxy, Lamiverre-A supplied by Nitto Electric Industry Co. Ltd. have been chosen. According to Kasen's code, it is defined as E1F42P-E/52.[3] The dimension of the specimen is 85 mm x 10 mm x 4 mm with a notch on each surface of which depth is 2 mm. The distance between the notches is 10 mm. The dimensions and shape of the specimen are shown in Fig. 1.

AE Measurements

Two AE sensors (PZT transducer) are attached on the different surface of the specimen through vacuum grease. The resonant frequency of the sensors is 140 kHz. The experiment is performed setting high pass filter 100 kHz and low pass filter 300 kHz. The AE sensors are located at the 10 mm distance from the notches. The position of the AE sensors is also presented in Fig. 1. The AE are monitored in ring down or event mode.

At the different load level, the applied force was unloaded and reloaded to check the Kaiser effect which confirmed the AE monitored was not the noise. At the each unloaded condition, the attenuation of the ultrasonic wave was measured to check the degree of interlaminar failure. In this case one of the AE sensor was used as a source of transmission pulse and another was used as a detector.

Interlaminar Shear Test

Interlaminar shear tests were made by Universal Testing Machine (Shimadzu Seisakusho Ltd) at RT and LNT. The tensile speed was 1.6 mm/min. The displacement of the specimen was measured by a clip gauge. In order to reduce the AE from gipped area, pin system has been chosen in stead of chuck. The AE sensors were also used to check the cavitation of liquid nitrogen in order to confirm the specimen was fully cooled down to LNT.

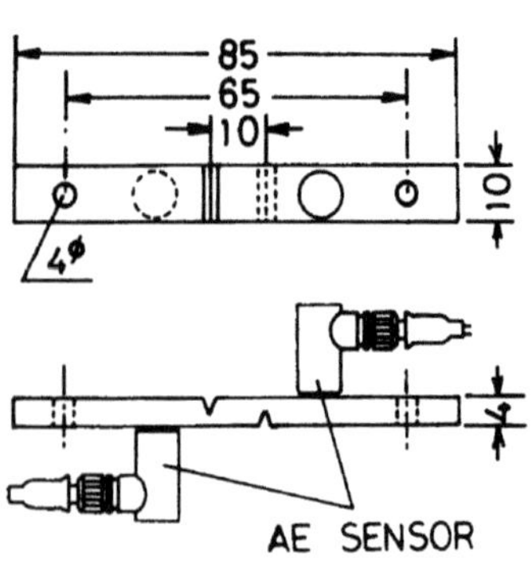

Fig. 1. Shape and dimensions of tested specimen. The arrangement of AE sensors are also shown. (Scale : mm)

Table 1. Interlaminar Shear Strength and Tensile Strength of Lamivvere A at Room and Liquid Nitrogen Temperature

	RT	LNT
Mean Shear Strength	2.55 (σ:0.10)	2.67 (0.23)
Maximum Shear Strength*	2.81	2.95
Tensile Strength	40.1 (3.6)	69.8 (5.5)

* Markhan et al. σ: standard deviation

RESULTS AND DISCUSSIONS

The tensile strength and mean interlaminar shear strength (ILSS) obtained at room (RT) and liquid nitrogen temperature (LNT) are presented in Table 1. The maximum ILSS calculated from mean ILSS is also shown in this table.[4] The tensile strength obtained at LNT is approximately 1.7 times as large as that obtained at RT. On the contrary the ILSS increases about 1.05 times. The tensile strength of composite materials depends on the strength of the glass fiber though the ILSS on the boundary strength between epoxy and fibers. The tensile strength of the glass fiber at LNT is approximately 1.6 times larger than that at RT.[5,6] The increase of strength of the present commposite material is thought good agreement with that of glass fibers.

Figure 2 shows the results of interlaminar shear tests obtained at RT and LNT. In this figure the cumulative AE count during the test in ring-down mode is also presented. As shown in table 1, the breaking load obtained at LNT is almost same as that obtained at RT. The AE counts measured at LNT, however, is approximately one order of magnitude larger than that at RT. The yielding is found on load-displacement curve at RT, though the serration at LNT. This shows that the fracture mode at RT is different from that at LNT.

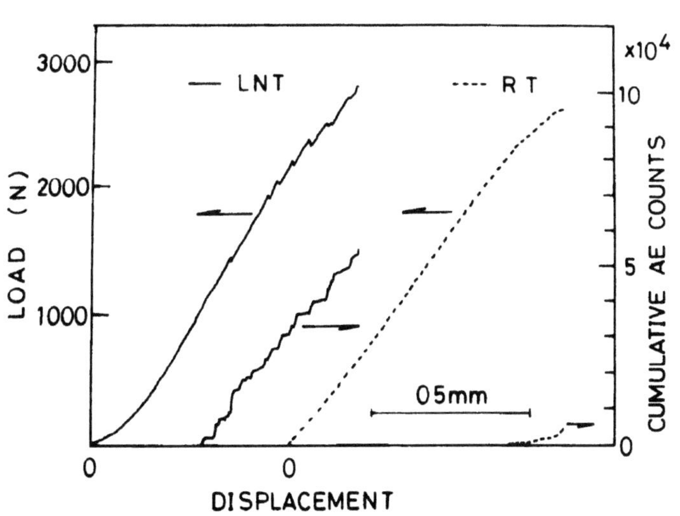

Fig. 2. Load-displacement curves and cumulative AE count obtained at RT and LNT.

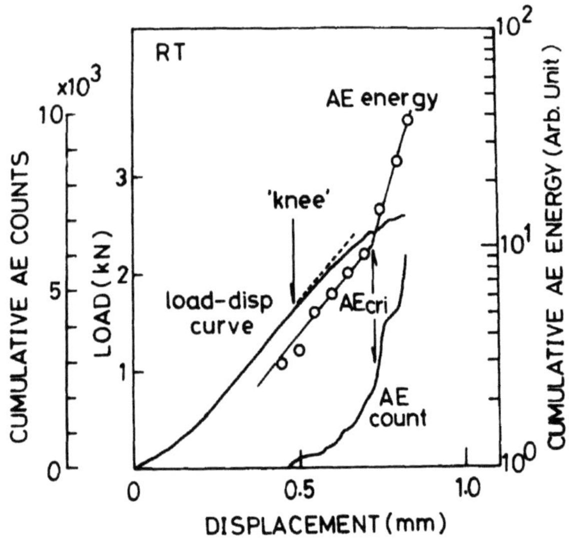

Fig. 3. Load-displacement cumulative AE count-displacement and cumulative AE energy-displacement curve obtained at RT.

The cumulative AE count-displacement curves are compared with the load-displacement curves. Figure 3 shows the load-displacement, cumulative AE count-displacement and cumulative AE energy-displacement curves obtained at RT. At RT the AE arises just when the "knee" point appears on the load-displacement curve (that is the load-displacement curve starts to bend).[7] The sudden increase of AE appears to correspond with the macroscopic yielding on the load-displacement curves. This point (named AE_{cri}) can be clarified in terms of cumulative AE energy. The AE energy is presented on a log scale in this figure. It can be seen that the AE energy rate suddenly increases just when the yielding starts on a load-displacement curve. The rapid increase of AE energy suggest the instable crack growth, that is, cracks propagate spontaneously.

Figure 4 shows the AE amplitude distributions both before and after AE_{cri} obtained at RT. Both the distributions are similar each other that is in low amplitude region straight line with negative slope and in high region a peak. The amplitude of the peaks agree with each other in both cases and suggests that the main cause of the AE is identical. The peak in the amplitude distributions are thought

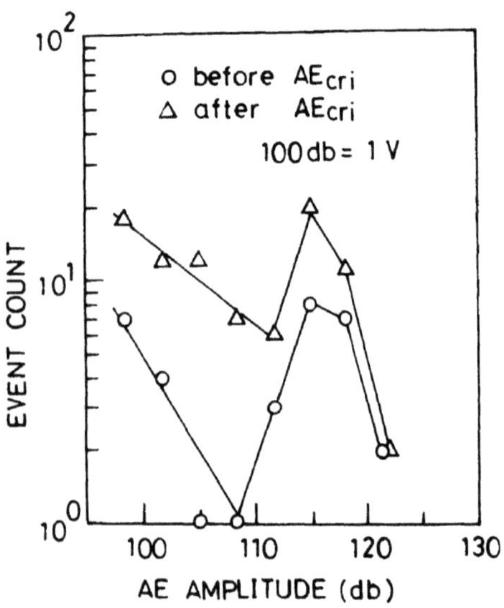

Fig. 4. AE amplitude distributions before and after AE_{cri} obtained at RT.

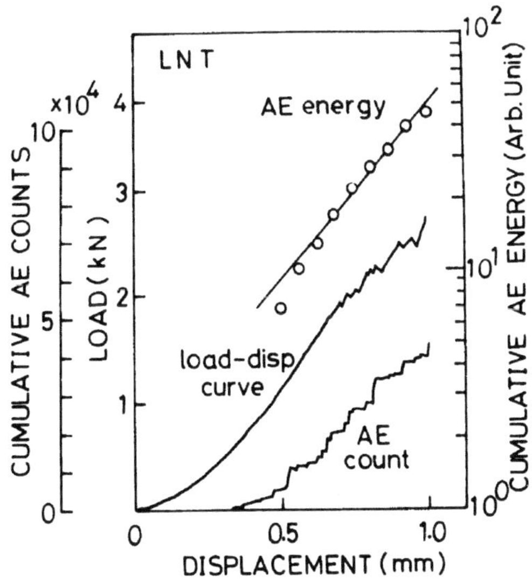

Fig. 5. Load-displacement, cumulative AE count-displacement and cumulative AE energy-displacement curve obtained at LNT

to be originated from the epoxy cracking and/or interlaminar failure considering the AE energy correspond to the peak.

The results obtained at LNT are presented in Fig 5. The "knee" point does not appear so clear as that at RT though small serration corresponding with AE stepwise increase even at low load level. At LNT the epoxy resin comes to be brittle and does not show flow. Because of this serration instead of yielding takes place. The change of AE energy is constant until fracture (increases exponentially) and hence sudden increase of instable crack does not continue as at RT. This means that instable crack growth is completed in short period. (mili seconds) and this corresponds to stepwise AE increase. Consequently cracks do not grow spontaneously that is the cracks do not grow until the applied force exceeds the previous level. On the contrary at RT when the applied force exceeds the AE_{cri} the specimen would break even if the deformation is stopped.

This asumption has been checked by stopping the deformation in the process of tensile test. Figure 6 shows the change of AE, displacement and applied load after the deformation was stopped at several load level. At approximately 70% of breaking load level (which estimated between "knee" and AE_{cri} point), the load was slightly relaxed and AE was emitted a few after the stop of deformation at RT. At this load level specimen did not break even after 15 minutes. Then the load level increased just around the AE_{cri} level. Even after the stop of the deformation, the load was relaxed and AE was emitted continuously and finally the specimen was fractured approximately 1 minutes after the stop of the deformation. On the contrary at LNT the specimen remained soundly at every load level (even at 95% of breaking load) and it is confirmed that the breaking load does not degradate. These behavior could be expected from the AE features.

Figure 7 shows the results of Kaiser effect and the AE (or ultrasonic wave) transmittance at RT and LNT. In the repetition of load-unload process, the AE is not emitted until the deformation (or the load) exceeds the previous displacement level (or load level). The phenomena is called Kaiser effect and this effect is used to confirm the detected AE is emitted from the material (i.e. not the noise). Both at RT and LNT Kaiser effect was found and hence the AE from the composite material was confirmed.

(a)

(b)

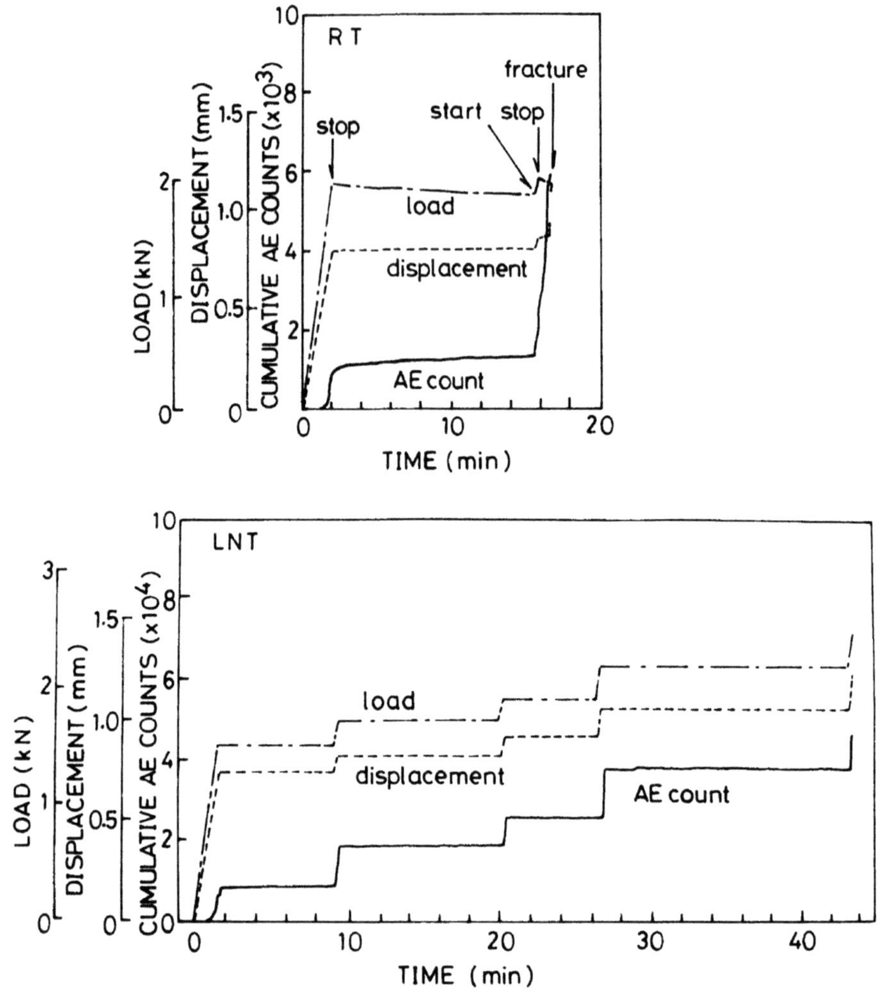

Fig. 6. Change of AE, displacement and applied force after the deformation stop at RT(a) and LNT(b).

In this figure the AE transmittance is plotted against the largest displacement (or load) ever experienced that is the displacement just before the unload. As expected from Fig. 1 the interlaminar cracks disturbs the propagation of AE and hence brings the decrease the transmittant AE amplitude. The degree of interlaminar failure can be estimated in terms of AE transmittance. The sudden decrease of AE transmittance which corresponds to yielding at RT and serration at LNT suggests the macroscopic crack growth. As expected from Fig. 2, at

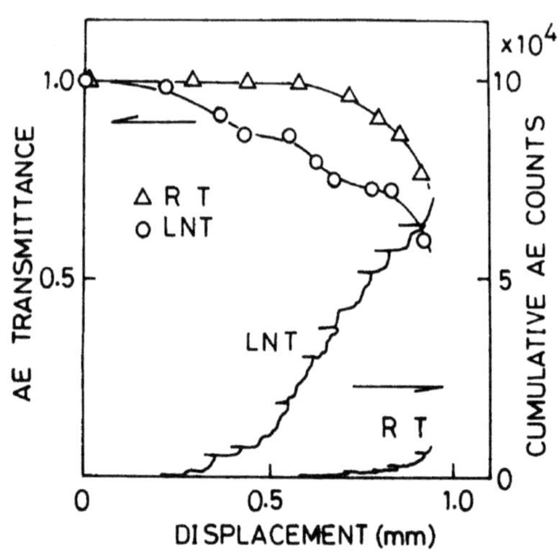

Fig. 7. Kaiser effect and AE transmittance at RT and LNT.

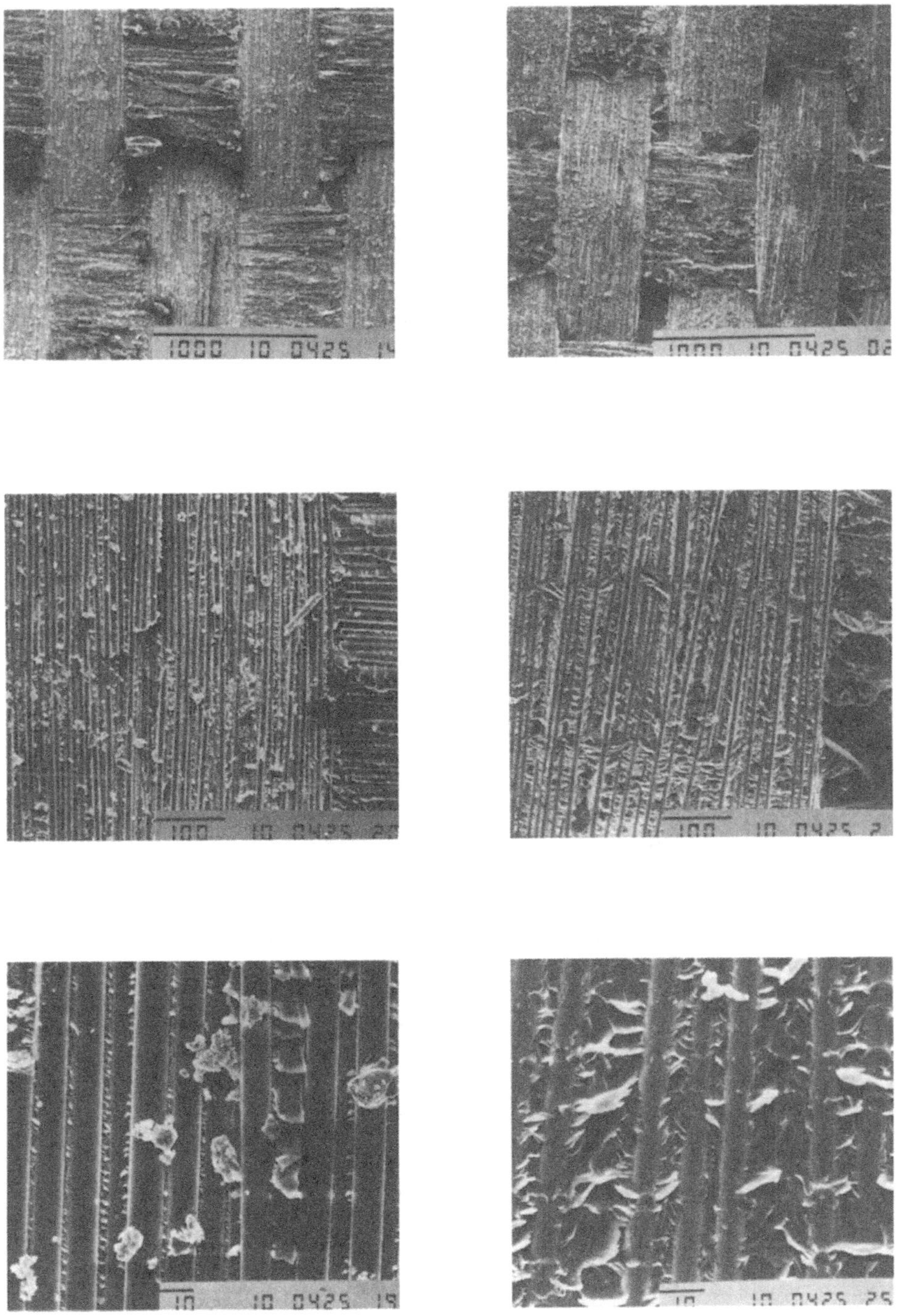

Fig. 8. SEM fracture surface characteristics obtained at RT (left row) and LNT (right row). The scale in the photographs is micron meter.

LNT the macroscopic cracks are introduced earlier stage than at RT. The decrease of transmittance agree well with the increase of AE count both at RT and LNT, and hence the AE detected are confirmed from the interlaminar failure.

In Fig. 8 SEM photographs of the fracture surface are presented. The specimen was pulled in the warp direction that is in these figures the vertical direction. The horizontally arraied texture (fill fibers) are disordered at RT during the interlaminar failure which suggests the epoxy matrix stick to the fibers sufficiently during the fracture.

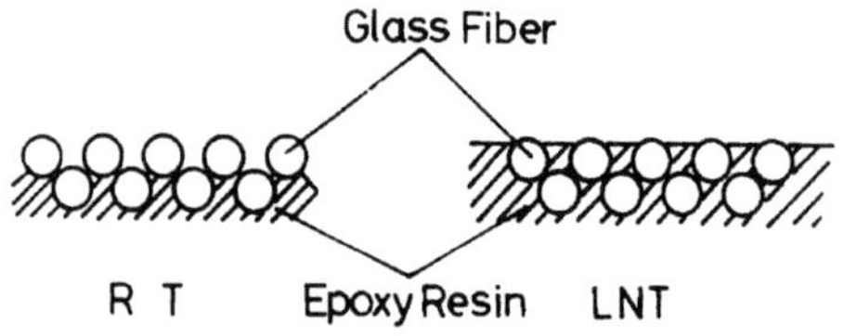

Fig. 9. Schematic illustration of fractured surfaces obtained at RT and LNT.

On the contrary at LNT the disordered texture does not appear so much as at RT which suggests the brittle fracture of epoxy at LNT. The fiber density of the fracture surface obtained at RT appears larger than that at LNT. Schematic illustration of fracture surfaces were shown in Fig. 9. At RT the fibers appear to be exposed on surface though at LNT they are buried in epoxy matrix. This is the reason that the fiber density of the fracture surface appears to be different. The epoxy matrix is peeled off from fibers at RT though fractured at LNT because at LNT the epoxy matrix comes to brittle. This is the reason that the disorder of fill fibers takes place at RT.

CONCLUSIONS

The interlaminar shear test have been made on glass reinforced epoxy and the AE during the test was monitored. The following conclusions were drawn.
(1) The fracture mode of interlaminar shear at LNT is different from that at RT though interlaminar shear strength obtained at LNT is almost same that obtained at RT. Serration and yielding takes place at LNT and RT, respectively. This is caused by the epoxy matrix change that is at RT epoxy matrix can flow plastically to some extent though at LNT the epoxy matrix comes to be brittle and cracks are easily introduced into epoxy matrix. This change can be detected by the AE features.
(2) The degree of interlaminar fracture can be estimated by the change of AE transmittance. The interlaminar fracture can be also detected by AE measurement technique.
(3) The spontaneous crack growth was detected as the rapid increase of AE energy.
(4) The amplitude distribution analysis of AE has the possibility as the identification method of fracture mode.

REFERENCES

1. T. Nishiura, S. Nishijima, K. Katagiri, and T. Okada, Mechanical behavior and acoustic emission of composite materials at cryogenic temperatures, in: "Proceedings of the International Cryogenic Materials Conference, Kobe, Japan," K. Tachikawa and A. Clark, eds., Buterworths, London (1982), pp. 278-281.
2. S. Nishijima, T. Okada, and S. Namba, Acoustic emission from composite materials at low temperatures, in: "Advances in Cryogenic Engineering," vol. 30, Plenum Press, New York (1984), pp. 25-32.
3. M. B. Kasen, Standardizing nonmetallic composite materials for cryogenic applications, in: "Nonmetallic Materials and composites at Low Temperatures-2," G. Hartwig and D. Evans, eds., Plenum Press, New York (1982), pp. 327-337.
4. M. F. Markhan and D. Dawson, Composites, 6 (1975), pp. 173-176.
5. D. L. Hollinger and H. T. Plant, The role of stress corrosion in glass fibers, in: "Proceedings of the 19th Annual Meeting of the Reinforced Plastics," The Society of the Plastic Industry, Inc. (1964) Section II-A

6. M. P. Hanson, Glass-, Boron-, and Graphite-Filament-Wound Resin Composite and Liners for Cryogenic Pressure Vessels, in: "Cryogenic Properties of Polymers," T. T. Serafini and J. L. Koenig, eds., Marcel Dekker Inc., (1967) pp. 29-43.
7. H. Fuchs, Machined gap laminates for components in heavy electrical engineering and their use at very low temperatures, in: "Nonmetallic Materials and Composites at Low Temperatures-2," G. Hartwig and D. Evans, eds., Plenum Press, New York (1982), pp. 355-366.

MECHANICAL PROPERTIES OF LOW THERMAL CONTRACTION GFRP

K. Fukushi, M. Nagai, Y.Kamata, and *K.Kadotani

Hitachi Research Laboratory, Hitachi, Ltd. Japan
*Yamazaki Works, Hitachi Chemical Co. Japan

ABSTRACT

Liquid helium duct spacers in large superconducting magnets must have high compressive strength to withstand large electromagnetic forces. They must also show a low thermal contraction, in the direction perpendicular to the laminations, to maintain windings stability at liquid helium temperature. To attain higher glass content or higher density, we adopted glass roving cloth with mineral fillers and unidirectional woven glass cloths. E-,S-, and R-glass cloths were examined as reinforcements and epoxy and bismaleimide triazine, as matrix resins.

The thermal contraction of GFRP with unidirectional woven cloths (cross ply) was nearly equal to that of stainless steel; compressive strength at cryogenic temperature was nearly 1200-1300 MPa. GFRP with glass roving cloths and fillers also showed low thermal contraction, however, their compressive strength was nearly 800-900 MPa.

INTRODUCTION

Conventional epoxy-matrix glass-reinforced insulating laminates (GFRP) have been used for many years in superconducting magnet construction. Initially, G-10 and G-11 of NEMA standards were widely used; later, Kasen et al.[1] developed cryogenic grade GFRP designated G-10CR and G-11CR. Recently, Hamelin[2] succeeded in developing high-density GFRP for the "TORE II" project. This GFRP has a linear thermal contraction equal to that of superconductors. Now, we have also succeeded in developing such a low thermal contraction GFRP which is described in this paper.

GFRP spacers for practical fusion reactors are exposed to up to 10^8 Gy (10^{10} rads) of radiation. As epoxy-matrix GFRP fails over 2×10^7 Gy at 5K[3], new materials using radiation resistant matrix resin are needed. Glass fibers for reinforcement should be boron-free for further radiation resistance. As such we first developed a high-density, low-thermal-contraction GFRP of epoxy-matrix. Then, we utilized polyimides and bismaleimide-triazine (BT) resins for the matrix and reduced their content to as low a level as possible. For reinforcements, R-, S-, and Eglass fabrics were employed. The final product was a high-density, low-thermal-contraction, and high-compressive-strength GFRP.

EPOXY GFRP

Component Materials

E-glass fabrics were used for the epoxy-matrix GFRP. Table 1 shows the component materials for the nine samples tested. Nos. 1 and No. 2 were standards; the former was G-11CR and the latter was "VL-E200". Nos. 3 to 9 were test samples. To decrease resin content, Nos. 3 to 5 adopted a unidirectional woven glass cloth in cross ply, and Nos. 6 to 9, glass roving cloth. Nos.7 and 9 used mineral fillers in combination with glass roving cloth. Nos. 3 to 7 with bisphenol A epoxy were manufactured by a prepreg-press method, the same as used for G-11CR and by Hamelin[2]. Nos. 2, 8 and 9 with cycloaliphatic epoxy were manufactured, by a vacuum-impregnation-pressure-molding method.

Density and Resin Content

Fig. 1 summarizes the relation between density and resin content of epoxy GFRP. G-11CR and VL-E200 had resin content of about 30 wt% and density of about 1.95 Mg/m^3. Our high-density samples had values of 16-24 wt% and 2.04-2.19 Mg/m^3. Nos. 7 and 9, with additional mineral fillers, had higher density than Nos. 6 and 8 with only glass roving cloth. Nos.3 to 5 were of the same component materials, however, they differed in curing pressure, which resulted in a different density and resin content. Nos. 5 and 7 had the highest density and lowest resin content.

Evaluation of Trial Samples

For superconducting magnet spacer usage, compressive strength (σ) as well as thermal contraction (α), both perpendicular to the laminations (subscript $\perp$), are the most important properties. Comp-

Table 1. Component materials of epoxy-glass laminates

No.	Reinforcements E-glass fiber	 Filler	Matrix resins
1*	Plain woven cloth		Bisphenol A epoxy Aromatic amine
2*	Plain woven cloth		Cycloaliphatic epoxy Acid anhydride
3	Unidirectional woven cloth (cross ply)		Bisphenol A epoxy Aromatic amine
4	Unidirectional woven cloth (cross ply)		Bisphenol A epoxy Aromatic amine
5	Unidirectional woven cloth (cross ply)		Bisphenol A epoxy Aromatic amine
6	Roving cloth		Bisphenol A epoxy Acid anhydride
7	Roving cloth	Alumina	Bisphenol A epoxy Acid anhydride
8	Roving cloth		Cycloaliphatic epoxy Acid anhydride
9	Roving cloth	Calcium carbonate	Cycloaliphatic epoxy Acid anhydride

*No. 1: G-11CR or "Spauldite SCR" from Spaulding Fibre Co.
No. 2: VL-E200 from Hitachi Chemical Co.

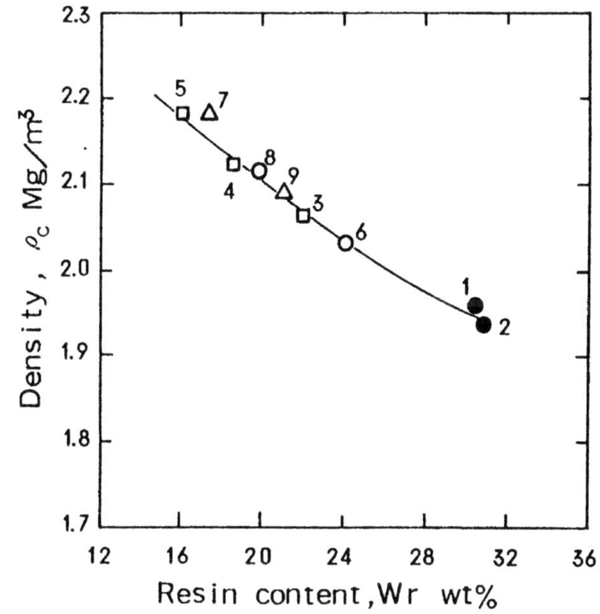

Fig. 1. Relationship between density and resin content of various epoxy-glass laminates.
□ :unidirectional woven cloth; ○:roving cloth; △:roving cloth with filler.

ressive strength $\sigma_\perp$ was measured at 77K (liquid nitrogen temperature) and thermal contraction $\alpha_\perp$, from 300K (room temperature) to 77K. Fig.2 evaluates the trial samples with regard to these properties; $\alpha_\perp$ should be lower and $\sigma_\perp$ higher. VL-E200 had better properties than G-11CR. Nos. 6 and 8 with glass roving cloth had lower $\alpha_\perp$ than VL-E200, and nearly the same $\sigma_\perp$ as VL-E200. Nos. 7 and 9 with additional mineral fillers, had lower $\sigma_\perp$ than VL-E200, and were nearly the same as for G-11CR. Nos. 3 to 5, with unidirectional woven cloth, showed excellent properties. No. 5 had an $\alpha_\perp$ as low as those of copper or stainless steel and $\sigma_\perp$ was over 1200MPa.

Analysis of Mechanical Properties

Tensile, flexural, interlaminar shear, as well as compressive properties were measured at 300, 77, and 4.2K (liquid helium temperature). Table 2 shows the ratios of various properties at 77K to 300K

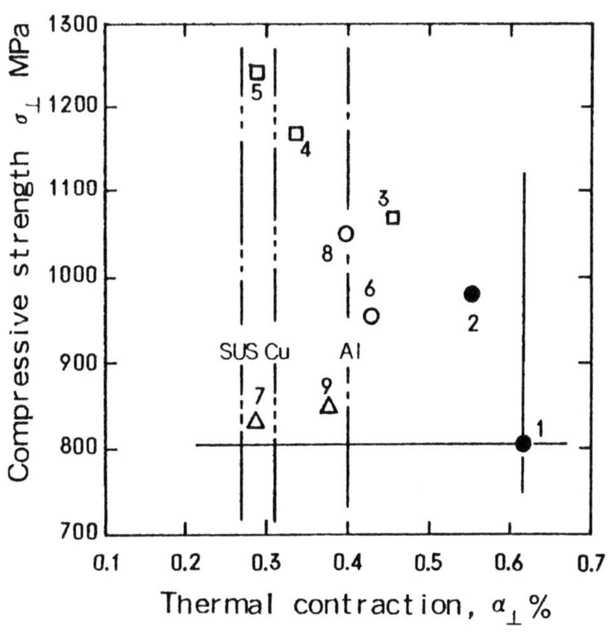

Fig. 2. Relationship between compressive strength (at 77K) and thermal contraction (from 300K to 77K) of various epoxy-glass laminates, both measured perpendicular to the laminations. Symbols: cf. Fig. 1.

Table 2. Ratios of various properties measured at different temperatures.

	77K/300K	4.2K/77K
Tensile strength	1.7	1.03
Flexural strength	1.7	1.04
Compressive strength	1.6	----
Flexural modulus	1.4	1.05
Tensile elongation	1.3	1.01
Flexural elongation	1.3	1.01
Interlaminar shear strength	1.2	1.05

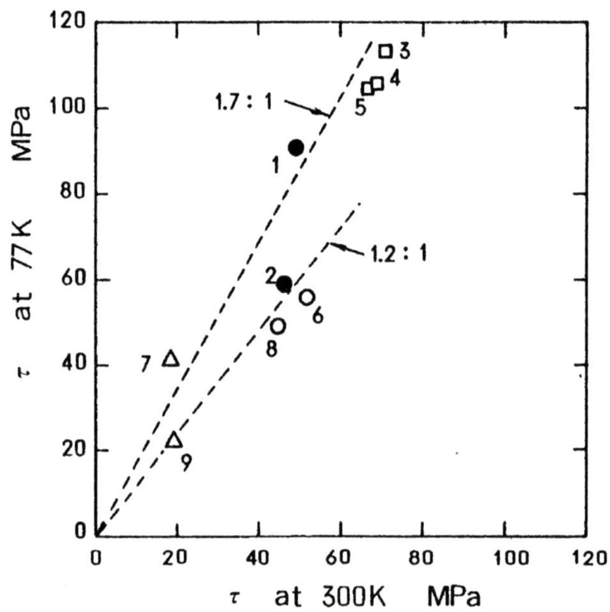

Fig. 3. Relationships between interlaminar shear strength at 77 and 300K for various epoxy-glass laminates. Symbols: cf. Fig. 1.

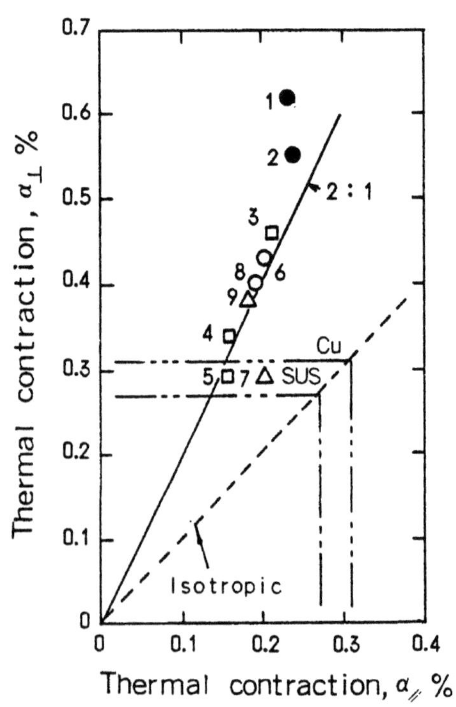

Fig. 4. Relationship between thermal contractions (from 300 to 77K) perpendicular and parallel to the laminations of various epoxy-glass laminates. Symbols: cf. Fig. 1.

and at 4.2K to 77K. The ratios at 4.2K to 77K were very small, however, those at 77K to 300K were large. Tensile, flexural, and compressive strength at 77K and at 4.2K were 60 to 70 % larger than at 300K; modulus was 40 % larger and elongation, 30 % larger.

Interlaminar shear strength (τ) was only 20 % larger. Fig. 3 shows the relation between τ at 77 and 300K. Interlaminar shear strength was measured by the short beam method. In the case of samples with higher τ, they burst by flexural stress, rather than shear stress. In Fig. 3, Nos. 1, 3 to 5, and 7 broke due to flexural stress and had a 77K/300K strength ratio of about 1.7, the same as that for flexural strength. However, Nos. 2, 6, 8, and 9 burst in the middle interlaminar surface due to shear stress. Thus, the ratio of the interlaminar shear strength (or the bonding stress) was assumed as 1.2.

Fig. 4 shows the relation between the thermal contractions perpendicular ($\alpha_{\perp}$) and parallel ($\alpha_{/\!/}$) to the laminations; $\alpha_{\perp}$ was nearly twice as larger as $\alpha_{/\!/}$. Thermal contraction of GFRP has been theoretically analyzed elsewhere[4].

RDIATION RESISTANT GFRP

Component Materials

The component materials of radiation resistant GFRP are shown in Table 3. No. 10 was the "Spaulrad", a polyimide-matrix E-glass plain woven cloth reinforced material. Nos. 11 to 16 were our trial samples. Bismaleimide-triazine (BT) resin (Mitsubishi Gas Chemical Co.) was used as matrix resin.

E-, S-, and R-glass fibers (Saint Gobain, France) were used as reinforcements. Nos. 11 and 14 were plain woven cloth; Nos. 12, 13, 15 and 16, were unidirectional woven cloth (cross ply).

Table 3. Component materials and properties of radiation resistant GFRP.

No.	Reinforcement	Density Mg/m^3	Resin content* Wt %
10**	E-glass plain woven cloth	2.04	26.5
11	E-glass plain woven cloth	2.02	25.1
12	E=glass unidirectional woven cloth (cross ply)	2.11	20.4
13	E-glass unidirectional woven cloth (cross ply)	2.16	17.7
14	S-glass plain woven cloth	1.94	26.7
15	R-glass unidirectional woven cloth (cross ply)	2.08	20.1
16	R-glass unidirectional woven cloth (cross ply)	2.19	14.7

*Matrix resin: No. 10: Polyimide; Others: Bismaleimide triazine
**No. 10: "Spaulrad" from Spaulding Fibre Co.

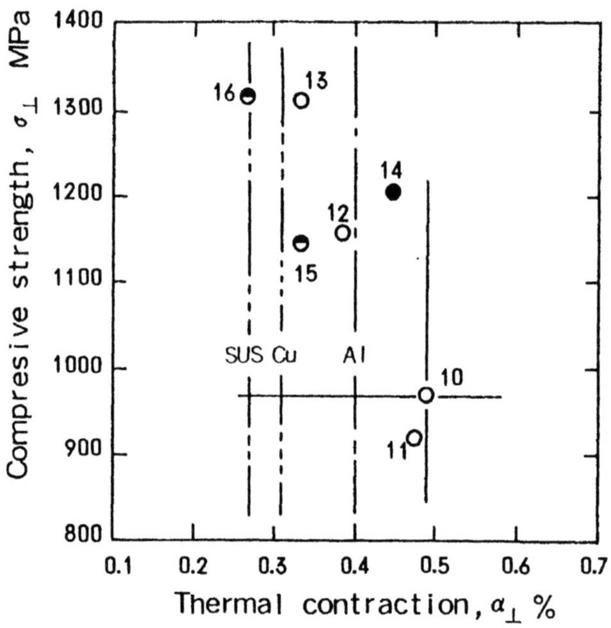

Fig. 5. Relationship between compressive strength (at 77K) and thermal contraction (from 300K to 77K) of various laminates listed in Table 3, both measured perpendicular to the laminations.
○ :E-glass; ● :S-glass
◓ :R-glass

Evaluation of the Properties

Evaluation of the trial samples is summarized in Fig. 5. Trial samples, with the exception of No. 11, had better properties than No. 10, "Spaulrad". No. 16 with unidirectional woven R-glass cloth showed the best properties; $\alpha_\perp$ was as low as that of stainless steel and $\sigma_\perp$ at 77K was over 1300 MPa.

Effect of Glass Composition

S- and R-glass fibers have different glass compositions from conventional E-glass. They are free of boron (or B_2O_3) and should show better resistance against neutron radiation. They should also have lower thermal contraction than E-glass. Fig. 6 shows the relationship between thermal contractions ($\alpha_\perp$ and $\alpha_{/\!/}$) and glass volume content. GFRP with S- and R-glass fibers showed lower thermal contractions than that with E-glass when glass content was the same.

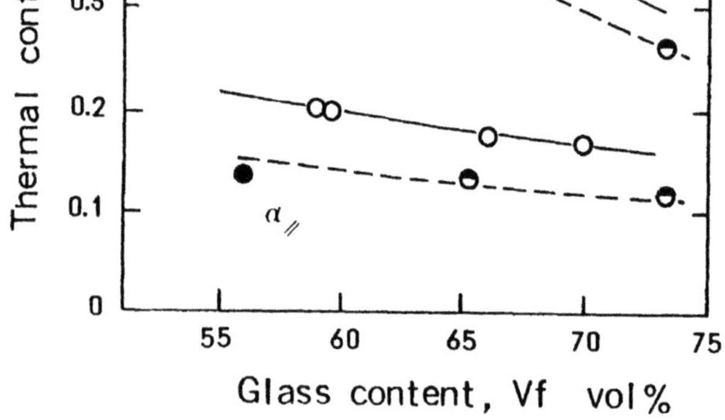

Fig. 6 Relationships between glass content and thermal contraction (from 300K to 77K) perpendicular to the laminations of various laminates listed in Table 3.
Symbols: cf. Fig. 5.

CNCLUSION

We have succeeded in developing high-density, low-thermal-contraction, and high compressive-strength GFRP for superconducting magnet usage. The materials were composed of cross-plied unidirectional woven cloth and epoxy or bismaleimide-triazine resin.

Their density was nearly 2.2 Mg/m^3, resin content was 16-17 wt %, thermal contraction ($\alpha_{\perp}$) was as low as that of stainless steel, and compressive strength ($\sigma_{\perp}$) at cryogenic temperatures was as high as 1300 MPa.

This study was performed in 1982, 1983 through Special Coordination Funds of the Science and Technology Agency of the Japanese Government.

REFERENCES

1. M. B. Kasen et al., Mechanical, electrical, and thermal characteristics of G-10CR and G-11CR glass-cloth/epoxy laminates between room temperature and 4K, "Advances in Cryogenic Engineering," 26, D-1, (1980)
2. J. Hamelin, Three-dimensional contraction and mechanical properties of glass-cloth-reinforced epoxy materials at cryogenic temperature, ibid., D-8, (1980)
3. R. R. Coltman et al., "The effect of a 100MGy(10 rads) gamma-ray dose at 5K on the strength of polyimide insulators," Report SSD No. 81-10, Oak Ridge National Laboratory, Oak ridge, Tennessee, (1981)
4. K. Kadotani and F. Aki, Low thermal contraction GFRP, Cryogenics, 23, 5:267-270 (1983)

ADHESIVE PROPERTIES AT LOW TEMPERATURE OF

EPOXY RESIN PRE-IMPREGNATED TAPE

D.Evans and J.T.Morgan

Rutherford Appleton Laboratory
Chilton, Didcot, England

INTRODUCTION

The work reported here was carried out as part of an investigation into the feasibility of building a large superconducting solenoid (approx. 5m diameter) using epoxide resin pre-impregnated glass fabric as coil insulant and bonding material.

The solenoid design specifies a cylindrical shell of aluminium alloy to contain the magnetic forces with a single layer of conductor wound into the inside of the shell. In order to minimise differential contraction between shell and windings, an aluminium clad superconductor was selected.

The programme of work was therefore limited to the bonding of substrates of aluminium and aluminium alloy. The investigation had four main objectives :-

i) Determine the pressures and temperatures necessary to achieve satisfactory bonding.

ii) Determine bond strengths in tension and shear at normal and low temperatures.

iii) Determine the effects of temperature cycling on bond strengths.

iv) Evaluate the effects of machined surface finish and treatment on the strength of the adhesive bond.

ADHESIVE TEST SPECIMENS

Where possible, the configuration of test specimens was chosen to be representative of the solenoid design. Two aspects of the use of insulation and bonding were considered :-

i) Insulation between shell and windings, (approx. 1mm thick)

ii) Conductor wrapping (approx. 0.25mm thick) to give interturn insulation and bonding.

A. Shear specimens

Three shear test specimens were used to investigate the two aspects of insulation.

The first type of specimen consisted of 4 aluminium bars, each approx. 120mm long and 25mm wide by 5mm thick. These were wrapped with 25mm wide pre-impregnated tape, stacked together and cured under load. The dimensions of the aluminium bars were similar to those proposed for the solenoid and the thickness of tape selected to give the correct interturn insulation thickness. Two modes of wrapping the tape were investigated, i.e. i) one tape wound half lap to give 2 layers of insulation thickness on each bar, and ii) 2 layers of tape wound butted, with the joints in each layer staggered.

The compressive loading applied to the stack of bars during cure was limited to that which it was considered could be achieved during solenoid manufacture.

The second type of shear specimen was used to investigate the bond strength of the shell insulation layer and the influence of surface finish on its adhesion. This consisted of two strips of aluminuim alloy with a 1mm thick layer of pre-impregnated glass fabric between cured at temperature under a compressive load of approx. one third of an atmosphere.

Both of these types of specimen were tested in three point loading with a short span, i.e. span to depth ratio approx. 3.5

Neither of these specimens introduced the differential thermal contraction effects which would be expected to arise in practise between the shell and the layer of stacked insulated conductors, and so a third type of shear specimen was devised. This consisted of a round bar onto which was wound and cured a 1mm thick layer of pre-impregnated tape. A stack of aluminium washers interleaved with glass fabric was bonded onto the outside, the proportion of composite between the washers was selected to give the same thermal contraction coefficient as that determined for a stack of insulated superconductors. The shear strength was determined from the load necessary to push out the central aluminium alloy bar.

B. Tensile specimens

These consisted of short (approx.40mm long) lengths of round (20mm diameter) aluminium alloy bar with one end machined flat and perpendicular to the axis and with the other end threaded for coupling to the testing machine. The flat ends of two of these were bonded together with the appropriate thickness of pre-impregnated tape, using a jig to hold the two ends square and concentric and to apply the correct compressive load on the bond during cure.

C. Surface preparation

Prior to bonding, all specimens were degreased in acetone, cleaned/etched in a 5% aqueous solution of a proprietary cleaner known as AC 51 and then rinsed with demineralised water before being dried in a warm air circulating oven.

D. Pre-impregnating resin system

A number of proprietary epoxide resin pre-impregnated glass fabrics and tapes capable of being cured at moderate temperatures (approx. 120°C) were subjected to preliminary screening tests. From these tests it appeared that one type of material was consistently superior to the others when bonded under relatively low pressures. This was the resin system examined in detail and was used for all of the tests in this report. It consists of a mixed solid and liquid Bisphenol A in the proportions of 1.5 : 1 with a latent acceletated Dicyandiamide (DICY) curing agent. The resin is applied to the glass fabric in solution and the solvent removed by passing through a drying tower, to give a rather tacky 'pre-preg'. Resin content is controlled at 43 ± 3 %. A dimpled interleaving film is used as this reduces the tendency to form 'pinholes' between the fibres when the interleaving is finally removed. 'Pinholes' could adversely affect the electrical performance of the laminate.

ANALYSIS OF SHORT BEAM SHEAR TEST RESULTS

If one assumes that the composite beam obeys the Engineeers Bending Theory, (i.e. normal planes remain normal and in plane), then the shear stress at an interface is given by :-

$$q = \frac{F\ A\ J}{I\ b}$$ where,

F = Shear force = Half applied load
A = Area above (beyond) interface
J = Centroid of A
b = Width of interface

However, experience has shown that where a high modulus material is connected by a series of low modulus interfaces, the structure deviates from this relationship. The limit is reached when the shear modulus of the adhesive approaches zero, and each layer acts independantly, so that the effective bending stiffness is the sum of the individual layers. The actual behaviour is somewhere between the two, but experience indicates that the system more closely follows the independant layer theory.

If there are n interfaces in the stack to carry the force F, then :-

$$q = \frac{F}{S\ b\ n}$$ where,

q = Average shear stress
F = Half failure load
S = Depth of one layer including insulation
b = specimen width
n = no. of layers

This relationship was used to calculate shear strengths in the short beam shear tests.

Table 1 shows the shear test results for type 1 specimens cured at 120°C under various compaction loadings. 25mm wide by 0.2mm thick glass fabric 'pre-preg', one layer wound half lap. The test was carried out at room temperature with and without thermal cycling down to -180°C.

TABLE 1

Compaction load g cm^{-2}	Shear strength MPa	After thermal cycling to -180°C MPa
420	31.5	35
420*	32	31.5
280	35.5	34.5
140	29	31

* Pre-preg stored 12 months at -18°C

The results for the shear tests using specimen type 2 with 5 layers of 0.2mm fabric between alloy bars with machined surface finishes of 100, 200 and 500 microinch are given in table 2. The results were obtained at room temperature, 77 K and 4.2 K using specimens approx. 24mm wide.

TABLE 2

Surface finish	Shear Strength MPa 298 K	77 K	4.2 K
500 L	28.5	46.0	53.8
500 T	27.7	-	-
200 T	27.2	42.4	43.2
100 T	26.7		

T denotes machine lines transverse to specimen length and L denotes longitudinal

Shear test results at low temperatures are given in table 3 for type 1 specimens compacted under a load of 420 g cm^{-2}.

TABLE 3

Tape thickness	Winding mode	Shear strength MPa 298 K	77 K	4.2 K
0.125mm	half lap	28.0	33.0	34.8
	double butt	-	34.5	32.3
	half lap*	18.2*	27.7	22.0

* Specimen prepared and compacted under load, then kept at room temp. not under load for 4 weeks, before being cured at 120°C under load.

* Specimen cycled 5 times to -180°C before testing.

The results from testing the type 3 specimens are shown in table 4.

TABLE 4

Test Temp.	Failure Load	Specimen Length	Shear Strength
	N	mm	MPa
R.T.*	35000	21.3	24.9
R.T.*	50000	32	23.7
77K	47000	21.3	33.5
77K	>100000	32	>47

* Specimens cycled 5 times to -180°C

In all cases the specimen failed at the bond between the 1mm layer of composite and the stacked washers rather than at the interface between the composite and the central alloy bar, (this bond, because of its smaller area withstanding a shear load of more than 26MPa at room temperature and 37MPa at 77K).

The results from the shear strength measurements indicate that satisfactory bonds may be achieved using low compaction pressures of approximately 280 g cm^{-2}. Within the range of machined finishes considered, no significant differences in bond strength were detected for the various surface finishes.

ANALYSIS OF TENSILE TEST RESULTS

The load required to fail the specimen was divided by the bond area to give the stress at failure. A low rate of test (0.5 mm min^{-1}) was selected to simulate a steadily increasing stress and to avoid the impact or 'snatch' effect. The specimen size of 20mm diameter was chosen so that failure would occur within the load limits of the testing equipment. Also, from previous experience with this type of specimen, 20mm diameter is sufficiently large so that the failure stress is not unduly influenced by edge effects. Care was taken during specimen preparation to ensure that the ends were aligned co-axially. A compressive load of 1.8Kg (560 g cm^{-2}) was applied to each specimen during cure.

TABLE 5

Tape thickness	Tensile Strength MPa 298 K	77 K	4.2 K
0.125mm	41.7	49.6	79.5
	38.2	54.1	56.7
	46.8	39.8	
	49.3		
	45.5		
0.09mm	35.6	94.5	89.1
	36.3	75.1	68.4
	36.3	67.8	
	45.2		
	36.9		

Note: The failure stress is for individual specimens containing 2 layers of the given thickness tape.

The bond strength in tension of this epoxide resin pre-impregnated glass fabric tape confirms its suitablity for use as a bonding material for aluminium and its alloys for use in cryogenic environments.

MDF CEMENTS FOR LOW TEMPERATURE APPLICATIONS

B.A.Hands[*], A.J.Smith[+*], G.W.Groves[+] and D.D.Double[+]

[*]Cryogenics Laboratory
Department of Engineering Science
University of Oxford, Oxford, England
[+]Department of Metallurgy and Science of Materials
University of Oxford, Oxford, England

ABSTRACT

A pilot study to investigate some low temperature properties of macro-defect free cements has been carried out. That based on Ordinary Portland Cement was initially unsuitable for use at cryogenic temperatures due to the development of thermal cracks on cooling. However, the addition of a latex enables strength to be maintained down to 4.2K. That made from High Alumina Cement is also found to be suitable for use down to 4.2K. Both types are non-porous, no leaks being detectable through a 2mm thick sheet when using a helium leak detector. Thermal contraction data has also been obtained.

Possible cryogenic applications include components for superconducting magnets, vacuum equipment, and containers for refrigerants.

INTRODUCTION

Cement is a material of a complex chemical composition which changes with time over a period of many years. Its brittleness is well known. It has recently been established[1] that the brittleness arises from both microscopic cracks and pores which are present in the cement when it is made: the colloidal pores influence the elastic modulus and fracture energy, while the 'crack-like' pores are the source of failure. The effect of the crack-like pores may be determined using the Griffith model of brittle fracture.

Recently, methods of controlling the porosity of cements (as opposed to concretes - mixtures of cement and small stones) have been devised[1]. These 'Macro-Defect Free' (MDF) cements have been patented by Imperial Chemical Industries[2], who now promote them under the name New Inorganic Materials (NIMs). Their characteristic feature is a typical maximum pore size of 15μm in the as-prepared condition, which produces a higher strength than ordinary cements, especially in tension. They are mixed using very little water, with the addition of a polymer to provide sufficient workability. The result is that there is no free water present, in contrast to the situation in ordinary cements and concretes. To ensure compaction, the mixing process must have a high shear component; twin-roll milling or extrusion are generally used. The consistency of the mix is similar to that of warm plasticene.

The pastes used in this work were made using a twin-roll mill. This took the form of two steel rollers of slightly different diameter, each rotating at the same speed, so that the paste was encouraged to remain on the larger roller. 'OPC' material was made using Ordinary Portland Cement with the addition of a 20% solution of a polyacrylamide in water. To make the paste, the polyacrylamide gel was applied to one roller, and the OPC powder was slowly added during rolling. Mixing was assisted by turning the mixture over and into the centre of the rollers using spatulas. The process must be completed within about 15 minutes, since due to the shear work applied, the cement heats up and dries out. The cement was then scraped off the rollers, rolled flat to thickness between polythene sheets and compressed to about 13 bars under a hydraulic press overnight. It was transferred to a humidity tank for one week, and left to dry out over silica gel for a further week. However, it was found that on cooling to 77K thermal cracks developed, so that the material had little strength in tension[3]. It was also found to be pervious to air, though to a lesser extent than ordinary cement mixes.

The addition of a latex rectified these deficiencies. This material, referred to here as 'OPC-L', was made in the same way, with the alternate addition of cement powder and latex solution to the mill until mixing was complete. The final composition by weight was 80.8% Ordinary Portland Cement (from Blue Circle Cement Company); 8.5% latex; 1.1% dry polyacrylamide; 9.6% water.

The second material used in this work was made from high alumina cement (HAC) type Secar 71 supplied by the Ciment Fondu Lafarge Company. The polymer used was a polyvinyl alcohol, Gohsenol Type KH17S produced by Nippon Gohsei. To make the paste, 7g of the polymer, which is in the form of a powder, was thoroughly mixed with 100g Secar 71. 14g of water was added, and the dry, crumbly mixture was passed through the twin-roll mill, rolled and pressed overnight as described above. The cement was then cured at 80°C for two hours, under light pressure to prevent warping.

Test specimens were cut from the appropriate hardened cement using a diamond saw, and lightly polished using fine emery paper.

POROSITY MEASUREMENTS

Porosity measurements were made on discs approximately 2cm in diameter and 2mm thick. No leak could be found through either material using a helium leak detector with a sensitivity of 1.7×10^{-14} Nm^3/s, giving an upper limit for the permeability to helium of 3×10^{-20} kg-mole $Pa^{-1}m^{-1}s^{-1}$. Attempts to measure a diffusion rate have so far been unsuccessful, since no evidence of diffusion could be found. For the permeability of an HAC mix to oxygen, ICI[4] quote values of 7×10^{-20} to 7×10^{-19} kg-mole $Pa^{-1}m^{-1}s^{-1}$.

MECHANICAL TESTING

Mechanical testing was carried out in bending, tests at 4.2K using a 4-point, dead loading machine, those at higher temperatures using a 3-point bending rig attached to an Instron testing machine. Since a larger volume of material is stressed in 4-point bending than in 3-point bending, and failure probably occurs from an internal flaw, the larger the stressed volume, the higher is the probability that a given flaw size will be present. Thus, since the strength will depend on the largest flaw present, one might expect the strength in 4-point bending to be lower than that in 3-point bending. A theoretical analysis using Weibull statistics[5] indicated that the 4-point bending strength is expected to be to 79% of the 3-point value. Tests at 77K and 300K on both machines confirmed this view. In this paper, all values obtained from 3-point bending have been reduced to the equivalent 4-point value by multiplying by 0.79. It should be noted that although the room temperature strength of the HAC cement is expected to be well in excess of 100MPa in 4-point bending, the values we obtained in this work were rather lower. The reason for this is not known.

The low temperature tests were conducted by immersion in an appropriate liquid: helium (4.2K), nitrogen (77K), a solution of solid carbon dioxide in ethyl alcohol (213K). At 333K, hot water was used. Although it is known that prolonged exposure to water weakens these cements, the test took only a few minutes and the environment was not expected to have a major effect.

Both cements increased in strength on cooling. The strength of the HAC mix rose steeply from 320K to about 200K and then remained relatively constant (Fig. 1). The strength of the OPC-L rose between 290 and about 200K, but was relatively constant outside this temperature range (Fig. 2). Preliminary data on the variation of Young's modulus with temperature is shown in Fig. 3.

Attempts to measure the Young's modulus of OPC-L at room temperature at high loads using the dead-loading machine met with no success, because the material underwent creep (Fig. 4). The HAC mix was also found to creep, but at a lesser rate. Conventional concretes and cements also creep, due to the migration of water. Since there is no free water in the MDF cements, the same mechanism cannot be cited here.

THERMAL CONTRACTION

Measurements of integrated thermal contraction were made relative to silica, using rods extruded from pastes made as described above. The results were:

	293-198K	293-77K
HAC	1.15×10^{-3}	1.53×10^{-3}
OPC-L	1.81×10^{-3}	3.89×10^{-3}
Dry concrete[6]	0.6×10^{-3}	1.0×10^{-3}

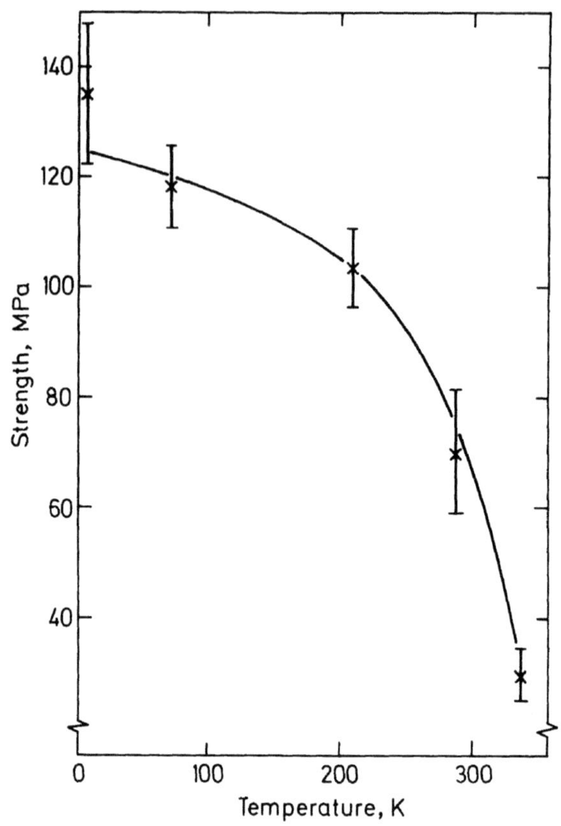

Fig. 1 Modulus of rupture of HAC MDF Cement

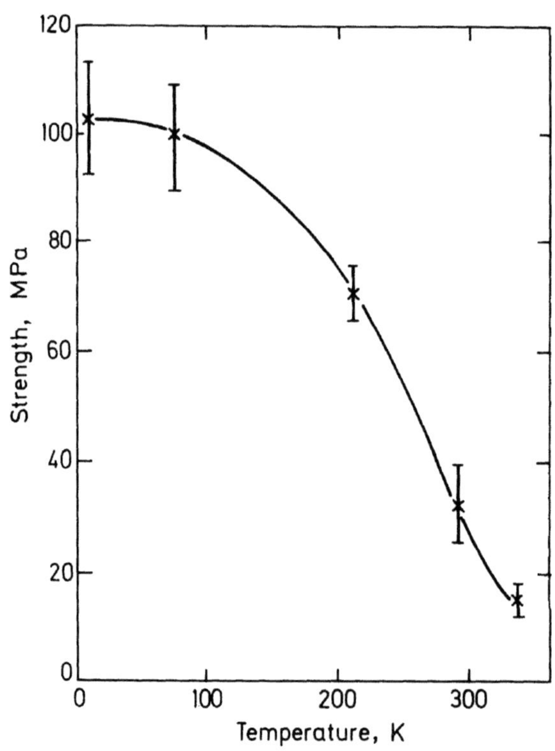

Fig. 2 Modulus of rupture of OPC-L MDF Cement

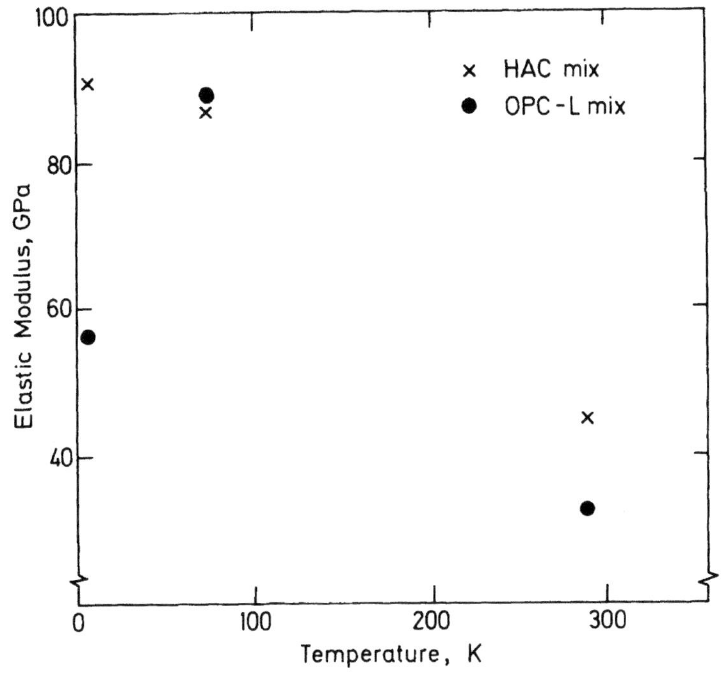

Fig. 3 Young's modulus as a function of temperature. The OPC-L data at 77K were obtained two months after manufacture.

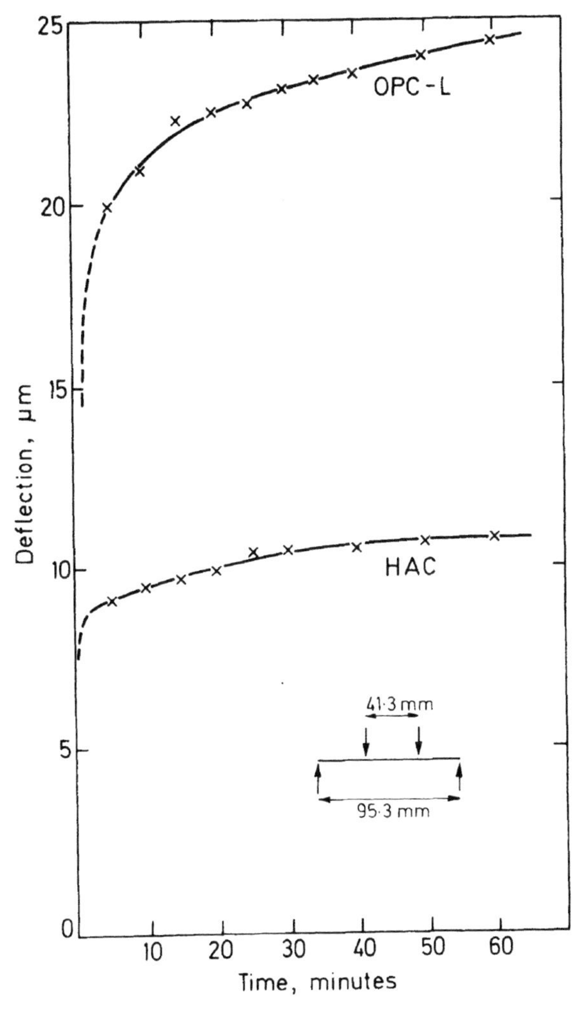

Fig. 4 Creep curves at room temperature. HAC: beam depth 3.0 mm; maximum fibre stress 18 MPa. OPC-L: beam depth 2.6 mm; maximum fibre stress 24 MPa.

At room temperature, the thermal expansion coefficient of a typical HAC mix has been given as $1\times10^{-5}/°C$.[4] The total contraction of silica is of the order of 10^{-5} over these temperature ranges[7], and so does not introduce a significant error.

DISCUSSION AND CONCLUSION

Since the MDF cements contain no free water, the change in properties with temperature cannot be due to the formation of ice, as with ordinary cements and concretes. We believe that a possible explanation is as follows.

Considering first the OPC-L material, the latex used was a carboxylated styrene-butadiene rubber, whose strength increases both with the fraction of COOH groups, and with the degree of vulcanisation (that is, the fraction of cross-links). It is known that carboxylated styrene-butadiene rubbers may be vulcanised at room temperature by calcium silicates, which constitute approximately 70% of Ordinary Portland Cement. The process requires the presence of divalent metal ions such as calcium and we suggest that the calcium ions in the cement cross-link the latex polymer chains via the carboxyl groups, hydrogen ions dissociating from the OH groups due to the alkaline environment prevalent in cement systems. If this speculation is correct, the latex is actually bonded to the cement grains, so that the material should perhaps be regarded as a cement-reinforced polymer, rather than as a cement material with the strength enhanced by the addition of latex. This idea is supported by the observation made using X-ray powder diffraction that, in contrast to ordinary cement, there is no detectable crystalline calcium hydroxide present[9].

The temperature dependence of the strength of OPC-L would then be directly attributable to the glass transition of the latex, which has a glass transition temperature of about 213K. Although cross-linking increases the glass transition temperature, the amount of cross-linking relative to the amount of latex is probably small, since only 0.5% by weight of the latex consists of carboxyl groups, so that we do not expect the glass transition temperature to be much altered. The polyacrylamide seems to be necessary as a rheological aid, but may play only a minor part in the strengthening mechanisms.

It is probable that similar processes operate in the HAC mix. The glass transition temperature of the KH17 Gohsenol as supplied is quoted as 328-333K. Thus one would expect the strong temperature dependence of the strength to occur at and above room temperature rather than below it as with OPC-L. Again, cross-linking is possible. KH17 Gohsenol is a polyvinyl acetate in which 80% of the acetate groups have been hydrolysed. As with OPC-L, it is postulated that hydrogen atoms dissociate from the polymer chain due to the high pH, and that metal ions cause both cross-linking and bonding between polymer and cement. This hypothesis is supported by some recent work which indicates that KH17 Gohsenol does indeed react chemically with the cement[8].

Thus the structure envisaged at present for both the HAC mix and the OPC-L is that of cement grains surrounded by a polymer film and bonded to it by a chemical reaction which also induces some cross-linking of the polymer. This is a structure which clearly invites comparison with fibre-reinforced materials, and explains the resistance of the materials to thermal cracking.

These views of the microstructures are supported by the creep data. Although ordinary cement and concrete is known to creep, over long periods the creep is usually attributed to the migration of residual water. In the cements examined here, the creep rate is probably attributable to creep of the polymer film surrounding the cement grains. In view of the relative glass transition temperatures of KH17 Gohsenol and the latex, the higher creep rate of OPC-L is not unexpected.

POSSIBLE USES

This study has established that MDF cements should be seriously considered for use in cryogenic equipment. They have a good failure strength in tension, although their lack of ductility may be a drawback. They may prove attractive in environments with a high magnetic field, for instance, as spacers in superconducting magnets. Their impermeability to helium also indicates uses for vacuum vessels and liquid containers, especially where low magnetic permeability and high electrical resistivity are required.

ACKNOWLEDGEMENTS

We are most grateful to David Evans of the Rutherford-Appleton Laboratory for allowing us to use his 4-point bending machine, and to Doverstrand Ltd. for their interest and help in supplying latices.

REFERENCES

1. K. Kendall, A.J. Howard and J.D. Birchall, Phil. Trans. R. Soc. Lond. A 310, 139-153 (1983).

2. European Patent Publications, nos. 0021682, 0038126, 0030408 (1981); no. 0055035 (1982).

3. B.A.Hands, A.Smith, G.W.Groves and D.D.Double, presented at 10th International Cryogenic Engineering Conference, Helsinki, Finland (August 1984).

4. ICI Mond Division Circular MHK/AP:109(1 February 1984).

5. R.W. Davidge, Phil. Trans. R. Soc. Lond. A310 113-123 (1983); Mechanical Behaviour of Ceramics (CUP, 1979)

6. A.L.Marshall, Cryogenics 22 555-565 (1982)

7. G.K.White, Experimental Techniques in Low-Temperature Physics (Clarendon Press, Oxford, 3rd. ed., 1979)

8. S.A. Brooks, S.A. Rodger, W.J. Sinclair, G.W. Groves and D.D. Double, paper to be presented at meeting of the Materials Research Society, Bosoton, Mass., U.S.A. (November 1984).

9. W.J.Sinclair, private communication.

PRESENT AND FUTURE APPLICATION OF NONMETALLIC MATERIALS IN CRYOGENIC TECHNOLOGY: Summary of the Panel Discussion

G. Bogner

Zentrale Forschung und Entwicklung, Siemens AG
Erlanger, F.R.G.

INTRODUCTION

This paper was prepared in order to facilitate the course of a panel discussion on the above topic. For simplicity the discussion is restricted to polymers - pure polymers, powder filled polymers and fibre reinforced polymers. Because of their outstanding properties, fibre reinforced polymers especially play an important role in cryogenic technology. In future this importance will further increase in accordance with the progress of knowledge in this field and the growing application of cryogenic technologies. Special attention is directed to the application of superconductivity whereas other application, like space research and cryogenic wind tunnels are not considered.

Today main epoxy resins filled with glass powder or fibres are used in the low temperature environment. But increasing attention is directed to fibre reinforced thermoplastic polymers, like polycarbonate (PC) and polysulfone (PSU). These materials exhibit some low temperature ductility and higher fracture strains at 4.2K than epoxy resins. Main problems with thermoplastic fibre composites arise from their fabrication technique and the fibre-matrix adhesion, which need further investigation. Finally carbon fibre-plastic composites are of increasing interest for low temperature applications.

GENERAL REVIEW ON THE APPLICATION OF PLASTICS AT LOW TEMPERATURES

Pure Plastics Pure materials are used as a thin layer varnish for the electrical insulation of superconducting magnets. Low viscosity epoxy resins are applied for vacuum impregnation of the windings of superconducting magnets in order to increase their mechanical rigidity. Because of the comparatively large thermal expansion coefficient ($\Delta l/_{e}$ (300K $\rightarrow$ 4.2K) $\simeq$ - 1.2%) of pure epoxy resins, large hollow spaces which are filled with resin during impregnation have to be avoided. Otherwise large thermal tensions are generated in these resin filled regions which can lead to heat generating cracks when additional high Lorentz forces are applied during charging the magnet. If the heat dissipated is large enough it may cause the magnet to go into the normal state. Therefore in general the conductors are wrapped with wrapped glass fibres or tapes and the space between layers is filled with thin glass fibre cloth. During impregnation the low viscosity resin penetrates into the glass fabrics, forming glass fibre reinforced epoxy with less tendency for generating cracks.

Often plastic tapes are used for the electrical insulation of superconductors. Because of its high mechanical strength and its excellent dielectric behaviour, wrapped Kapton tape is a preferred candidate. Wrapped tape insulation has also been suggested for the electrical insulation of superconducting high power cables. Favoured materials are oriented polypropylene and oriented polyethylene because of their high Young's modulus and low dielectric losses.

Storage vessels for cryogenic liquids (He, H_2, N_2) are generally equipped with a multilayer insulation to reduce the heat input by thermal radiation. This "Superinsulation" consists of thin ($\approx$ 5 μm) plastic foils, eg. polyethylene terephtalate, coated with a thin reflective layer of aluminium.

Powder Filled Plastics The main reasons for fusing powder filled plastics in a cryogenic environment are enhancement of mechanical strength and especially the opportunity of adapting the thermal expansion coefficients and the thermal conductivities of such composites within certain limits. Glass powder filled composites are frequently used as fitting and supporting pieces in superconducting devices. The force coupling between different structural parts in large superconducting magnet systems, eg. winding package and He-vessel, often is accomplished by filled resins. An interesting method of realising a positive coupling is the usage of thin walled stainless steel cushions, blown up under pressure by filled polymers.

Instead of glass powder, Al_2O_3-powder is often used as filler in composites. Besides the adaption of thermal contraction (300K $\rightarrow$ 4.2K), the thermal conductivity of such composites is considerably increased, compared with the unfilled polymers. Aluminium oxide powder-resin composites are applied for the impregnation (by wet winding)of superconducting coils, remarkably improving the mechanical strength and the cooling conditions of such windings. These composites are also often used for adapting and coupling pieces in superconducting devices. Comparatively high thermal conductivities can be achieved by mixtures of resins and pure metal powders, like Cu and Al.

Glass Fibre Reinforced Plastics (GFRP) Fibre reinforced plastics play an important role in technical applications at cryogenic temperatures. They combine the advantages of high mechanical strength and low thermal conductivity. They are especially suited for low thermal loss supports or suspensions of cryogenic liquid vessels in ambient temperature vacuum mantles of cryostats. This kind of suspension is frequently used for the mounting of superconducting coils (plus He-vessel) within their ambient temperature vacuum tank.

For applications where a high mechanical strength and a low weight are of great importance, eg. for space applications, cryostats completely manufactured from GFRP are used. This is also true for applications where cryostats are exposed to strong magnetic fields at power frequency or to small high frequency fields. In this case the diminishing electrical conductivity of the plastic parts of these cryostats prevent the generation of eddy currents, which otherwise would cause intolerably high electrical losses and/or undesirable disturbances of the applied field pattern. If such a cryostat accommodates a superconducting coil, the coil former also has to be manufactured from a plastic material.

For the construction of superconducting magnets resin preimpregnated glass fabrics or tapes ("Prepregs") are frequently applied. With the proper treatment quite significant shear strengths can be achieved with these materials. But in many cases, eg. bending magnets for large ring accelerators, their contribution to the mechanical rigidity of the winding package is small but they facilitate handling during construction.

Vacuum Tightness and Irradiation Resistance of Polymers With all polymers a serious problem arises from their large helium permeability. This is due to the fact that the molecular diameter of helium is smaller than the inter-molecular clearance of the macromolecules of the polymers. Therefore the walls of He-vessels, which are vacuum insulated, have to be coated with thin metallic layers (eg. Al) which have a considerable lower He-porosity than plastics. The additional use of He absorbing getters is recommended.

Plastics used in superconducting magnet systems for particle accelerators and especially for fusion reactors will be eventually exposed to dangerous irradiation dose rates (eg. 10^{17} - 10^{18} n/cm^2 per lifetime). This irradiation may have a negative influence on the mechanical and electrical characteristics or these plastics. Epoxy resins with inorganic fillers, which were irradiated with corresponding dose rates, showed a different degradation of their characteristics, depending on the filling material, the type of resin and cure. From these experiments it is anticipated that with the expected dose rates no critical degradation of the material properties will occur. But further investigations are necessary.

PRESENT AND FUTURE APPLICATION IN SUPERCONDUCTING DEVICES

Magnets for High Energy Physics High energy physics is one of the areas where superconducting magnets are being applied on a large scale. In the past large powerful detector magnets and beam handling magnets have been constructed and operated. Recently (1983) a large 1000 GeV colliding beam accelerator (TEVATRON) was put into operation at the US Fermi Laboratory. This machine consists of about 1000 magnets with a total length of 6 km. A very similar accelerator (HERA) will be built in the second half of the eighties at DESY, Hamburg, Germany. Its magnetic length also measures about 6 km. About 4 tons of Kapton will be used in this accelerator (tapes for the electrical conductor insulation, foils for protection layers between windings and mechanical support structure). In addition nearly 2 tons of prepreg-material and more than 1 ton of GFRP (adapting and distance pieces) will be employed.

The next generation of large beam colliding accelerators is already in the design stage. In the US and Europe studies are under way on machines with energies of 10 to 20 Tera eV (Large Hadron Collider at CERN, Geneva, Super Superconducting Collider in the US). The construction and operation of these accelerators is provided for the nineties. Their size will be many times that of the machines mentioned above. The magnetic length of the CERN-LHC is planned to be about 27 km, with a corresponding demand of plastic material, which is a factor four to five higher than for HERA.

Magnets for Magnetic Resonance (MR) Imaging and Spectroscopy
MR-imaging and spectroscopy for medical diagnosis are the field of the first large scale industrial application of superconducting magnets. Worldwide about 200 magnets will be manufactured in 1985 and 500 to 1000 units per year are predicted for the future. The magnets are of considerable size, about 1m clear bore at ambient temperature, outer diameter 2 to 2.5m, length 2 to 3m. The field strengths vary from 0.5T to 4T.

With these magnets great attention is paid to low thermal losses in order to achieve long time intervals between He-refillings. For this reason the suspension of the 4.2K cold mass (up to several tons) within the 300K vacuum tank is accomplished by GFRP members. In addition resins are applied for conductor insulation, coil impregnation and auxiliary components. Some manufacturers use GFRP for the inner cryostat tube which accommodates the patient. By this measure disturbances of the pulsed gradient field by eddy currents shall be avoided. For this reason also the formers for the gradient coils and compensation coils at room temperature are fabricated from fibre glass reinforced plastics. The total amount of plastics incorporated eg. in a 2 T-magnet is 100 to 200 kg.

Magnets for Industrial Processes It is expected that industrial processes is the area of the next large scale application of superconducting magnets. First magnetic separation has to be mentioned. Superconducting separators will cut energy costs and improve the efficiency. The performance and advantages of superconducting separators have already been demonstrated by some prototype plants. The size and magnetic flux densities of such magnets are of the same order of magnitude as with MR-magnets, ie. the amount of plastics necessary per unit is similar to that of MR-magnets.

The application of high magnetic fields during the growth of semi-conductor single crystals also is of great interest. Considerable improvements in the cryostat quality can be achieved by this method.

Superconducting magnets have also been proposed as bending magnets for small and compact electron-storage rings, which are provided to be used as synchrotron radiation sources for x-ray lithography in the manufacturing process of very highly integrated semi-conductor devices.

Magnets for Controlled Thermonuclear Fusion Fusion reactors which are based on magnetic confinement need very large superconducting magnet installations, which will consume hundreds of tons of plastic material. But it is expected that the first commercial demonstration power plant will not go into operation before the end of this century.

Today some scaled down experimental superconducting magnet systems are under construction, eg. a Tokamak system at the US Oak Ridge National Laboratory, a mirror machine at the US Lawrence Livermore, National Laboratory and a Tokamak installation in France (TORE SUPRA II). One of the six magnets of the Oak Ridge project has been manufactured by Siemens. In total about 2500 kg of epoxy-glass composites have been used in this magnet (impregnation of the winding package, positive couplings, adapting pieces and electrical insulation).

Today studies are under way for the next generation of experimental fusion installations, which will have magnet extensions similar to the size of future commercial plants. In Europe it is planned to start construction of such a machine (NET = next European Tokamak) in the nineties.

Superconducting Generators Today superconducting generators are predominantly developed in Japan, USSR and Federal Republic of Germany. The present rotor designs are conservative they use austenitic steels as rotor materials and the superconducting field winding is imbedded in slots machined into the rotor body. Polymers are only used for the electrical insulation of the conductor and the winding spacers, and for cooling channels.

More advanced designs also suggested the use of fibre reinforced polymers for the rotor body. Because of the comparatively low mass density of these materials such designs would lead to smaller mechanical loads on the rotor body by centrifugal forces. But other difficulties have prevented the realization of these proposals until now.

Further Large Scale Applications of Superconductivity In the past remarkable efforts were made to introduce superconducting components with higher efficiencies, higher capacities and lower capital costs into the existing power system. Superconducting high power cables, with electrical plastic tape insulation were investigated and their practicability has been demonstrated with test cables of considerable length (100 m). But because of economic reasons superconducting cables will not be in use before the year 2000.

Large superconducting dipole magnets have been built within programs for developing MHD-(magneto hydrodynamic) power plants. Meanwhile these programs have been cancelled, because of technical problems with the channel materials together with economical problems.

Studies on superconductive magnetic energy storage (SMES) for load levelling in electrical utilities have been carried out especially in Japan, US and Germany. Magnet systems of enormous size (several hundred meters diameter) have to be mounted in narrow underground tunnels in bedrock, which serves as a warm mechanical support for the tremendous magnetic forces of the magnet system. The force transmission from the cold parts to the warm bedrock is accomplished by epoxy-glass fibre composites. For a storage system with a capacity of 4GWh some 1000 tons of epoxy glass fibre composites would be necessary. However the probability of realizing such a large storage system is small.

With respect to a new high speed ground transportation system based on superconducting magnets for levitation, guidance and propulsion, the technical feasibility of such a system was demonstrated by large successfully operated test facilities in the Federal Republic of Germany and especially in Japan. Whilst the German project was finished in 1979, a final decision on the continuation of the Japanese project has not yet been taken.

The last named applications would require large amounts of non-metallic materials. But today a prognosis on their chances of realization is impossible.

Superconducting Quantum Interference Devices - SQUIDS SQUIDS are used as magnetometers with an extremely high sensitivity (10^{-14} Tesla $Hz^{-1/2}$). They are especially suited for measurements of biomagnetism = magnetocardiography and encephalography. For fully utilizing the high sensitivity of these sensors, their cryostats have to fulfill serious requirements, ie. non-magnetic and non-metallic materials have to be applied. Until now glass-epoxy composites have been used, but the use of thermoplastic polymers is also considered. For the glass tissue a material with an acceptable content of iron oxide impurities has to be chosen. A serious problem arises from the He-permeability of plastic fibre composites, which is solved by coating the vessels exposed to He by a thin aluminium layer. In addition the generation of magnetic fields by the thermoconductive and thermoreflective parts of the cryostats has to be avoided.

SQUID-cryostats are already commercially available. Larger quantities will be needed when the new medical diagnostic method has received its clinical approval.

RESUME

Only applications of non-metallic materials (=plastics) in the area of applied superconductivity have been considered in this paper. It could be demonstrated that even with this restriction already quite a remarkable amount of plastics are applied in this field. Today the main areas are: magnets for high energy physics, magnetic resonance imaging and magnets for industrial processing (separation etc). The demand for these applications is continuously increasing. Additional demand can be expected for systems which today are not yet finally developed, eg. magnet systems for fusion reactors, or which are not applied at present because of economic restrictions, eg. superconducting cables.

MATERIALS INDEX

SUBJECT INDEX

GPSR Compliance

The European Union's (EU) General Product Safety Regulation (GPSR) is a set of rules that requires consumer products to be safe and our obligations to ensure this.

If you have any concerns about our products, you can contact us on ProductSafety@springernature.com

In case Publisher is established outside the EU, the EU authorized representative is:

Springer Nature Customer Service Center GmbH
Europaplatz 3
69115 Heidelberg, Germany

Batch number: 10684218

Printed by Printforce, the Netherlands